RESEARCH OPPORTUNITIES IN GEOGRAPHY AT THE U.S. GEOLOGICAL SURVEY

Committee on Research Priorities in Geography at the
U.S. Geological Survey

Board on Earth Sciences and Resources
Division on Earth and Life Studies

NATIONAL RESEARCH COUNCIL
OF THE NATIONAL ACADEMIES

THE NATIONAL ACADEMIES PRESS
Washington, D.C.
www.nap.edu

THE NATIONAL ACADEMIES PRESS   500 Fifth Street, N.W., Washington, DC 20001

NOTICE: The project that is the subject of this report was approved by the Governing Board of the National Research Council, whose members are drawn from the councils of the National Academy of Sciences, the National Academy of Engineering, and the Institute of Medicine. The members of the committee responsible for the report were chosen for their special competences and with regard for appropriate balance.

This study was supported by Cooperative Agreement No. 00HQAG0217 between the National Academy of Sciences and U.S. Geological Survey, Department of the Interior. Any opinions, findings, conclusions, or recommendations expressed in this publication are those of the author(s) and do not necessarily reflect the views of the organizations or agencies that provided support for the project.

International Standard Book Number is 0-309-08516-0

Additional copies of this report are available from the National Academies Press, 500 Fifth Street, N.W., Lockbox 285, Washington, DC 20055; (800) 624-6242 or (202) 334-3313 (in the Washington metropolitan area); Internet, http://www.nap.edu

Front and back covers: The Geographic Face of the Nation (land cover), 2001. SOURCE: U.S. Geological Survey.

Cover designed by: Van Nguyen

Copyright 2002 by the National Academy of Sciences. All rights reserved.

Printed in the United States of America

THE NATIONAL ACADEMIES
Advisers to the Nation on Science, Engineering, and Medicine

The **National Academy of Sciences** is a private, nonprofit, self-perpetuating society of distinguished scholars engaged in scientific and engineering research, dedicated to the furtherance of science and technology and to their use for the general welfare. Upon the authority of the charter granted to it by the Congress in 1863, the Academy has a mandate that requires it to advise the federal government on scientific and technical matters. Dr. Bruce M. Alberts is president of the National Academy of Sciences.

The **National Academy of Engineering** was established in 1964, under the charter of the National Academy of Sciences, as a parallel organization of outstanding engineers. It is autonomous in its administration and in the selection of its members, sharing with the National Academy of Sciences the responsibility for advising the federal government. The National Academy of Engineering also sponsors engineering programs aimed at meeting national needs, encourages education and research, and recognizes the superior achievements of engineers. Dr. Wm. A. Wulf is president of the National Academy of Engineering.

The **Institute of Medicine** was established in 1970 by the National Academy of Sciences to secure the services of eminent members of appropriate professions in the examination of policy matters pertaining to the health of the public. The Institute acts under the responsibility given to the National Academy of Sciences by its congressional charter to be an adviser to the federal government and, upon its own initiative, to identify issues of medical care, research, and education. Dr. Harvey V. Fineberg is president of the Institute of Medicine.

The **National Research Council** was organized by the National Academy of Sciences in 1916 to associate the broad community of science and technology with the Academy's purposes of furthering knowledge and advising the federal government. Functioning in accordance with general policies determined by the Academy, the Council has become the principal operating agency of both the National Academy of Sciences and the National Academy of Engineering in providing services to the government, the public, and the scientific and engineering communities. The Council is administered jointly by both Academies and the Institute of Medicine. Dr. Bruce M. Alberts and Dr. Wm. A. Wulf are chairman and vice chairman, respectively, of the National Research Council.

www.national-academies.org

COMMITTEE ON RESEARCH PRIORITIES IN GEOGRAPHY AT THE U.S. GEOLOGICAL SURVEY

WILLIAM L. GRAF, *Chair,* University of South Carolina, Columbia
BARBARA P. BUTTENFIELD, University of Colorado, Boulder
CAROL P. HARDEN, University of Tennessee, Knoxville
JOHN R. JENSEN, University of South Carolina, Columbia
GEORGE P. MALANSON, University of Iowa, Iowa City
PATRICIA F. MCDOWELL, University of Oregon, Eugene
SARA McLAFFERTY, University of Illinois, Urbana-Champaign
RISA I. PALM, University of North Carolina, Chapel Hill
NORBERT P. PSUTY, Rutgers University, New Brunswick, New Jersey
HENRY J. VAUX, JR., University of California, Riverside

National Research Council Staff

ANTHONY R. DE SOUZA, Director, Board on Earth Sciences and Resources
LISA M. VANDEMARK, Study Director
MONICA R. LIPSCOMB, Research Assistant
TERESIA K. WILMORE, Project Assistant

COMMITTEE ON GEOGRAPHY

BILLIE L. TURNER II, *Chair,* Clark University, Worcester, Massachusetts
BERNARD O. BAUER, University of Southern California, Los Angeles
RUTH S. DEFRIES, University of Maryland, College Park
ROGER M. DOWNS, Pennsylvania State University, University Park
MICHAEL F. GOODCHILD, University of California, Santa Barbara
SUSAN HANSON, Clark University, Worcester, Massachusetts
SARA L. McLAFFERTY, University of Illinois, Urbana-Champaign
ELLEN S. MOSLEY-THOMPSON, Ohio State University, Columbus
ERIC S. SHEPPARD, University of Minnesota, Minneapolis

National Research Council Staff

KRISTEN L. KRAPF, Program Officer
MONICA R. LIPSCOMB, Research Assistant
VERNA J. BOWEN, Administrative Associate

BOARD ON EARTH SCIENCES AND RESOURCES

RAYMOND JEANLOZ, *Chair,* University of California, Berkeley
JILL BANFIELD, University of California, Berkeley
STEVEN R. BOHLEN, Joint Oceanographic Institutions, Washington, D.C.
VICKI J. COWART, Colorado Geological Survey, Denver
DAVID L. DILCHER, University of Florida, Gainesville
ADAM M. DZIEWONSKI, Harvard University, Cambridge, Massachusetts
WILLIAM L. GRAF, University of South Carolina, Columbia
RHEA GRAHAM, New Mexico Interstate Stream Commission, Albuquerque
GEORGE M. HORNBERGER, University of Virginia, Charlottesville
DIANNE R. NIELSON, Utah Department of Environmental Quality, Salt Lake City
MARK SCHAEFER, NatureServe, Arlington, Virginia
BILLIE L. TURNER II, Clark University, Worcester, Massachusetts
THOMAS J. WILBANKS, Oak Ridge National Laboratory, Tennessee

National Research Council Staff

ANTHONY R. DE SOUZA, Director
TAMARA L. DICKINSON, Senior Program Officer
DAVID A. FEARY, Senior Program Officer
ANNE M. LINN, Senior Program Officer
PAUL M. CUTLER, Program Officer
KRISTEN L. KRAPF, Program Officer
KERI H. MOORE, Program Officer
LISA M. VANDEMARK, Program Officer
YVONNE P. FORSBERGH, Research Assistant
MONICA R. LIPSCOMB, Research Assistant
EILEEN McTAGUE, Research Assistant
VERNA J. BOWEN, Administrative Associate
JENNIFER T. ESTEP, Administrative Associate
RADHIKA CHARI, Senior Project Assistant
KAREN L. IMHOF, Senior Project Assistant
SHANNON L. RUDDY, Senior Project Assistant
TERESIA K. WILMORE, Project Assistant
WINFIELD SWANSON, Editor

Acknowledgments

This report has been reviewed by individuals chosen for their diverse perspectives and technical expertise, in accordance with procedures approved by the NRC's Report Review Committee. The purpose of this independent review is to provide candid and critical comments that will assist the authors and the NRC in making their published report as sound as possible and to ensure that the report meets institutional standards for objectivity, evidence, and responsiveness to the study charge. The content of the review comments and draft manuscript remain confidential to protect the integrity of the deliberative process. We wish to thank the following individuals for their participation in the review of this report:

Brian J. L. Berry, University of Texas, Dallas
Vicki Cowart, Colorado Geological Survey, Denver
Michael F. Goodchild, University of California, Santa Barbara
Judy Olson, Michigan State University, East Lansing
Dallas L. Peck, USGS *emeritus,* Reston, Virginia
Stanley W. Trimble, University of California, Los Angeles
Thomas J. Wilbanks, Oak Ridge National Laboratory, Oak Ridge, Tennessee

Although the individuals listed above have provided many constructive comments and suggestions, they were not asked to endorse the conclusions or recommendations, nor did they see the final report before its release. The review of this report was overseen by Freeman Gilbert, University of California, San Diego. Appointed by the National Research Council, he was responsible for making certain that an independent examination of this report was carried out in accordance with institutional procedures and that all review comments were carefully considered. Responsibility for the final content of this report rests entirely with the authoring committee and the NRC.

Preface

Geography plays an increasingly important role in education, science, policy making, and government operations. The inclusion of geography in public school curricula has produced a generation of young people who are better trained in the subject than in previous years (NRC, 1997). Geography has also become a more prominent component of science as described by the recent National Research Council (NRC) report *Rediscovering Geography: New Relevance for Science and Society.* New recognition by decision makers of the local to global context of the issues they face has brought geography into play at every level. The refinement of geographic technology, including geographic information systems and remote sensing, has given powerful tools to analysts and decision makers that actuate fundamental geographic principles in experimental research and applied problem solving.

In the twenty-first century the U.S. Geological Survey (USGS) should take advantage of geographic theory and techniques, including electronic databases, new maps (to replace the original topographic maps) and application of remote sensing to resource and hazards management. In this spirit, John Kelmelis, Chief Scientist for Geography at the USGS, requested that the NRC conduct a study for multiple audiences about geography at the USGS. The Committee on Research Priorities in Geography at the U.S. Geological Survey was charged to identify high-priority basic and applied research opportunities in geography as they relate to the science goals and responsibilities of the USGS. Generally, the committee was asked to address the following areas of concern to the Geography Discipline (the portion of the agency once known as the National Mapping Division) of the USGS: (1) the role of the USGS in advancing the state of geographic knowledge of the discipline (geography, cartography, and geographic information sciences), (2) the role of the USGS in improving the understanding of the dynamic connections between the land surface and human interactions with it, (3) the role of the USGS in maintaining and enhancing the tools and methods for conducting and applying geographic research, and (4) the role of the USGS in bridging the gap between geographic science, policy making, and management. The committee included members from a range of sub-disciplines in geography, including geomorphology, policy for public land and water, cartography, geographic information systems, hydrology, remote sensing, biogeography, landscape ecology, hazards, urban systems, social geography, and economics.

The committee met four times in Washington, D.C., and visited the USGS headquarters and its Eastern Regional Center in Reston, Virginia. Sub-groups of the committee also visited the USGS Regional Center in Denver, Colorado, as well as the USGS EROS Data Center in Sioux Falls, South Dakota. As input to the study the committee reviewed the scientific literature and data, and materials from the Survey and other federal agencies. The committee also greatly benefited from discussions with the National Aeronautics and Space Administration (NASA) and the National Imagery and Mapping Agency (NIMA).

The committee acknowledges the many individuals who gave briefings or provided input during the course of the study (see Appendix B). The committee was most fortunate to work in a supportive NRC environment. Lisa Vandemark managed the many details that surfaced along the way. Monica Lipscomb was exceptionally helpful in this project, particularly in the assembly of the final manuscript. Anthony de Souza, director of the Board on Earth Sciences, contributed important ideas and insights, and he was a marvelous collaborator in refining the report.

William L. Graf
Chair

Contents

EXECUTIVE SUMMARY 1

1 INTRODUCTION 13
Geography and the Changing USGS, 14
Science at the USGS, 17
USGS Vision and Mission, 19
The Geography Discipline at the USGS, 23
Study and Report, 23

2 GEOGRAPHY AT THE USGS 25
Modern Geography, 25
History of Geography and Geographers at the USGS, 31
Present Geography Contributions to the USGS, 37
Summary, 46

3 PRIORITIES FOR MAINTAINING AND ENHANCING
SPATIAL DATA MANAGEMENT 47
Primary Priorities, 48
Secondary Priorities, 55
Summary, 63

4 PRIORITIES FOR GEOGRAPHIC INFORMATION
SCIENCE 65
Primary Priorities, 65
Secondary Priorities, 77
Summary, 82

5	RESEARCH ON LAND-SURFACE AND SOCIETY INTERACTIONS Primary Priorities, 84 Secondary Priorities, 91 Summary, 97	83
6	CONCLUSIONS AND RECOMMENDATIONS Advancing the General State of Knowledge in the Discipline, 101 Understanding the Dynamics of the Land Surface-Human Activities Connection, 101 Maintaining and Enhancing Geographic Tools and Methods, 102 Bridging the Gap between Science, Policy Making, and Management, 105 Summary, 106	99

REFERENCES 107

APPENDIXES

A	BIOGRAPHICAL SKETCHES OF COMMITTEE MEMBERS	115
B	ORAL AND WRITTEN CONTRIBUTORS	121
C	ACRONYMS	125
D	LETTER FROM P. LYNN SCARLETT	129

Executive Summary

Imagine it is the year 2011, a decade after the U.S. Geological Survey (USGS) redefined priorities in its geography research and increased emphasis on the "Critical Zone," the earth's surface and near-surface environment that sustains nearly all terrestrial life (NRC, 2001a, p. 35; Sidebar 1-1). A decade after the September 11, 2001, terrorist attacks in New York and Washington, D.C., the USGS is a fundamental contributor to the nation's homeland security. By increasing the research emphasis within the Geography Discipline (formerly the National Mapping Division) and enhancing the interaction between the Survey and its research partners, the USGS has assumed a major role in applying geography to problem solving.

When in 2011 terrorists introduced deadly bacteria into the water supply of several U.S. cities, the Survey used geographic information systems (GIS) and spatial analysis to predict the pathways of the contaminants. The Survey then mapped the distribution of likely casualties by splicing social science data with hydrologic data.

Using the virtually real-time *National Map*, the USGS specified efficient mass evacuation routes and provided city maps so complete that they included rivers and streams, as well as water mains and pipes. Based on the Survey's past experience and research into responses to natural hazards, such as volcanoes and earthquakes, the USGS identified likely bottlenecks in the delivery of emergency medical, food, and shelter supplies. The Survey's regional specialists, integrators of information about the natural and social systems of the areas attacked, were able to advise federal, state, tribal, and local officials in managing the massive dislocations and in reclaiming the damaged water supplies. After the biological attacks the Survey provided geographic information and new knowledge about spatial processes that improved decision-support systems. These computer programs allowed decision makers and individual citizens to refine their response to the threat using a series of "what if" scenarios.

USGS VISION AND MISSION

Such a role for the USGS would be a reality if the Survey capitalized on its opportunities. The Survey is undergoing reform, by redefining itself as an organization capable of supplying natural science products that are globally recognized as credible, objective, and relevant to society's needs. The USGS is focusing its activities on the "Critical Zone," the earth's surface and near-surface environments, where humans most directly interact with the natural system (NRC, 2001a, p. 35; see Sidebar 1-1). Its challenging mission is to provide reliable scientific information to:

- describe and understand the earth;
- minimize loss of life and property from natural disasters;
- manage water, biological, energy and mineral resources; and
- enhance and protect quality of life.

To achieve this mission the USGS has organized itself into four disciplines—the Geography Discipline, Geology Discipline, Water Discipline, and Biology Discipline. Each Discipline emphasizes a regional structure, resulting in a greater focus on geographic integration of its activities. The Geography Discipline assumes a position of prominence at a time when U.S. geography is re-emerging with newly recognized relevance to science and society. The juxtaposition of disciplinary change (throughout geography) with organizational change (throughout USGS) creates an unusual opportunity for geography to fulfill the Survey's mission in new and innovative ways.

NRC COMMITTEE CHARGE

In 2000, to maximize the benefits of internal reform in the USGS and external change in geography, the USGS invited the National Research Council (NRC) to form a committee of experts to advise on issues related to research priorities in geography throughout the Survey. The Committee on Research Priorities in Geography at the U.S. Geological Survey was asked to address, for multiple audiences, the society's need for geographic research and the appropriate federal research role. Specifically, the committee was charged to consider the following areas of concern for the Geography Discipline:

- The role of the USGS in advancing the state of knowledge in the discipline (geography, cartography, and GISciences);
- The role of the USGS in improving understanding of the dynamic connections between the land surface and human interaction with it;
- The role of the USGS in maintaining and enhancing tools and methods for conducting and applying geographic research; and
- The role of the USGS in bridging the gap between science, policy making, and management.

The overall goal of the study was to provide a fresh perspective and guidance to the Geography Discipline about its future research and strategic directions.

The importance of defining the role of the USGS in these broad geography-associated areas is in maximizing the Survey's potential to serve the nation's needs. The nation is entering a digital era when business activities, policy decisions at all levels, and citizens' individual choices rely on accurate data and clear understanding of the dynamics of the nation's geography. The location and distribution of resources and the people who depend upon them, the patterns of the natural and built landscapes, and the processes at the interface of nature and society are essential geographic issues that face the United States and the USGS.

THE USGS AND GEOGRAPHY

Geography has a venerable past at the USGS, a transitional present status, and a promising future. Before 1900, geography was a central part of the Survey's activities and it spent half its annual budget on items related to geography and mapping. By the twentieth century, however, geology and hydrology had become the foci of Survey research, and geography lapsed into the supporting function of map production. At the end of the twentieth century, biology was assigned a strong research role. Outside the Survey, geography has become a quantitative spatial science, a discipline concerned with the interface between nature and society. Inside the USGS, however, geography is almost solely a technical function, producing maps and imagery, rather than acting as an engine for research.

The USGS's scientific composition—geography, geology, hydrology, and biology—is comprised of personnel who operate within a distinctly regional framework in Reston, Virginia; Denver, Colorado; and Menlo Park, California. Yet, with 1,274 government employees, more than 369 contract employees, and an annual budget of about $133 million the Geography Discipline continues primarily as an organization devoted to mapping and imagery support. The presence of significant numbers of researchers in USGS disciplines is essential for the fulfillment of their mission, which emphasizes science. Disciplines with strong research and large numbers of Ph.D.-trained investigators in the specialty of the discipline create an atmosphere of creative inquiry where there is a culture of challenging questions, rigorous analysis, and thoughtful agendas. At present the Geology Discipline includes about 500 Earth science Ph.D. holders; the Biology Discipline has about 400 Ph.D. bioscientists; and the Water Discipline has about 200 Ph.D. hydroscientists. There are 10 Ph.D. geographers in the Geography Discipline. A small number of geography Ph.D. holders work in USGS disciplines other than the Geography Discipline.

Four themes run through the vision of this committee for the future of geography at the Survey:

1. The Geography Discipline should engage in scientific research.
2. The geographic research throughout the USGS should provide integrative science for investigations of the Critical Zone [i.e., the earth's surface and near-surface environment that sustains nearly all terrestrial life (NRC, 2001a, p. 35); see Sidebar 1-1].
3. The Geography Discipline should develop partnerships within the Survey and with the field of geography outside the Survey.
4. Geography should develop a long-term core research agenda that includes several projects of the magnitude of *The National Map*.

After assessing the geographic initiatives undertaken by several other agencies, the committee considers that the research agenda for geography presented in this report is appropriate for a federal natural science agency and complements the work undertaken by other federal natural science agencies. The need to generate new knowledge, tools, and methods through research in geography is most easily seen by comparison with the other disciplines at the Survey. In the Geology Discipline personnel and their clients expect not only outstanding geologic maps but also scientific analysis and explanations of geologic processes. In the Water Discipline members and clients expect accurate data regarding water resources, as well as insightful analysis of surface water and groundwater processes. In the Biology Discipline members and clients expect census data concerning endangered species, in addition to new understandings of the dynamics of ecosystems supporting those species. In the same way, agency members and their clients should expect the world's best spatial databases from the USGS, plus explanations and predictions for the spatial processes that those databases depict. Closer collaboration among the Survey's four disciplines will improve the quality of science research at the USGS.

CONCLUSIONS AND RECOMMENDATIONS

Towards a productive, useful future for the Geography Discipline, the NRC committee makes the following conclusions and recommendations, organized according to the charge.

The General State of Geographic Knowledge

The role of the USGS in advancing the state of knowledge of geography is to take a leadership position in a few critical areas of the field. The Survey is uniquely suited to provide leadership in GIScience (i.e., the research necessary to develop and sustain investigations into spatial phenomena, including the computer science behind geographic information systems, the physical science behind remote sensing, and the spatial statistics behind geographic analysis. It can also contribute in other primary areas by development of a suite of general geographic applications that focus on nature-society interactions.

Conclusion: Currently the USGS's influence is weak in advancing the state of knowledge in general geography (i.e., geographic research other than GIScience) because Survey personnel conducting such research are not sufficiently engaged with geographers outside of the Survey.

Recommendation: To advance the state of knowledge in geography in general the USGS should strengthen its connections to the scientific community outside the Survey. These connections will be improved if Survey personnel participate in national geographic organizations and present USGS geographic research at professional geography meetings and in professional journals.

Conclusion: The USGS's influence is weak in advancing the state of knowledge in general geography because geographers at the Survey are limited to cartographic, geographic information systems (GIS), and remote sensing specialties, largely at the technical level.

Recommendation: The USGS should expand its capabilities in geography beyond the activities of cartographic technicians to include leading-edge geographic research in GIScience, spatial analysis, and nature-society interactions.

The Dynamics of the Land Surface–Human Activities Connection

The USGS, in advancing the state of knowledge in geography as a general discipline, specifically contributes information related to the Critical Zone, that part of the earth system that is at or near the surface and that is the home of humanity. The Survey's mission statement recognizes special responsibilities in natural disasters that result from the interaction of society with nature. It is appropriate for a federal natural science and information agency to take a national leadership role in research aimed at improving the understanding of this interaction as it relates to hazards.

Conclusion: The USGS manages large amounts of data to assess processes at the nature-society interface and provides a supporting mechanism for responses to natural disasters. Even though the fact that the academic field of geography is a significant contributor to the understanding of environmental processes and natural hazards, the Survey does not contribute greatly to the understanding of the vital connection between nature and society through scientific research focused on hazards.

Recommendation: The USGS should continue to exercise national leadership in applied hazards research (including natural, technical, and security hazards) to improve the nation's explanatory, predictive, and response capabilities. To meet national needs, however, it is incumbent on the Survey to undertake basic research on environmental processes, hazards, and vulnerability, and to include the expertise of geographers and social scientists from within the Survey or through cooperative agreements.

Conclusion: The USGS manages and provides a variety of basic data for the nation's responses to natural and technical hazards. These data and methods of analysis are also applicable to issues related to homeland security, a subject that has many data and research similarities to investigations of natural and technical hazards.

Recommendation: The USGS should implement a homeland security support system founded on the general principles used by the Survey for dealing with natural hazards.

Geographic Techniques

Traditionally the USGS has maintained and enhanced tools and methods in geography to fulfill the Survey's role as one of the nation's primary sources of spatial data. The committee sees great potential in the Geography Discipline for an expanded role in the areas of research related to spatial data, as well as research to support cutting-edge geographic products and an information base such as *The National Map*. *The National Map* as a database product and an information base is an attainable goal by 2010, but some of the basic knowledge needed to create it (and other spatial data products) is not yet available. In assessing the role of the USGS in maintaining and enhancing tools and methods, the committee distinguishes between issues related to data from a more general consideration of GIScience.

Data Research

Conclusion: The USGS manages a national treasure of historic data ranging from maps and remotely-sensed imagery to long-term data collected from biologic, hydrologic, and geologic systems. These historic data are not artifacts valuable only for their curiosity. Rather, they indicate long-term trends in natural systems and baseline measures to assess human influences. Historic data allow the interpretation of present data, but use of the historic information is restricted by several unsolved problems related to access, processing, and analysis.

Recommendation: The USGS should develop projects focused on historic data to address basic geographic research questions related to the accuracy, availability, quality, and scale issues for historical spatial data.

Conclusion: The terrorist attacks of September 11, 2001, raise issues regarding data security, especially for data the USGS manages, including imagery, maps, and water supply data. The Survey's data management responsibilities are conflicting. On one hand, one of the Survey's purposes is to make these data widely available; on the other, the federal government has a responsibility to protect data that might be used against the nation. At the USGS four associate directors determine which data to make available within their own Discipline. Because only general guidelines are available, the four associate directors' restrictions could be inconsistent.

Recommendation: A uniform security policy for spatial data should be developed, and the associate directors should serve as advisors to a single

USGS decision maker. To make as much data available as possible the policy should clearly outline how the mission of the Survey and the security of the nation should be balanced in making decisions for data management.

Conclusion: Although Congress has designated the USGS as the clearinghouse agency for spatial data, other Department of the Interior (DOI) bureaus and federal agencies create and use spatial data. The underlying problem is a lack of integration among these geospatial databases (those databases with locational identifiers attached to data entries), which does not serve the scientific and public good. Addressing this problem requires research, standards, and the application of integrating methods.

Recommendation: The USGS is ideally suited to be the lead agency in providing and managing spatial data, and the federal government should make available resources commensurate with the level of the task. The USGS should play a leading and facilitating role in shaping national policy on geospatial data and developing an interoperable capability that will make it a primary access point for integrated geospatial data in the Department of the Interior and other federal agencies.

GIScience Research

Conclusion: *The National Map* is a bold vision for the future of the Geography Discipline, with the spatial database of the same name being its most prominent product. Without question the digital era has made the paper topographic map series obsolete for many applications, but *The National Map* will not become a reality with our present level of knowledge about the tools and methods needed to create the product.

Recommendation: Given the importance of *The National Map* to the information economy of the future, and the need for further supportive research to accomplish *The National Map,* the Geography Discipline's programs—Cooperative Topographic Mapping, Land Remote Sensing, and Geographic Analysis and Monitoring—should receive a level of funding commensurate with the task.

Conclusion: Construction and maintenance of *The National Map* will require a variety of databases, but some databases are of exceptional priority if *The National Map* is to succeed. These high-priority datasets will require emphasis in funding and support.

Recommendation: Because of their importance in supporting *The National Map*, the following datasets should be assigned the highest priority in distribution of resources and in establishing and improving interagency exchanges:

- orthorectified imagery;
- digital elevation data;
- land cover data;
- biogeographic data;
- hydrographic data;
- transportation feature data; and
- geographic place names.

Conclusion: The USGS is effective in creating and managing spatial data, but its role in GIScience is limited and does not include cutting-edge research in geographic information systems or the analysis of the data that the Survey provides to others. The Survey has a weak research program in geographic science related to the discipline's tools and methods.

Recommendation: To achieve the vision and mission of the USGS the Survey should improve its contributions to geographic knowledge, tools, and techniques by developing the internal capability to address the high-priority subjects of:

- resolution and scale;
- delivery of vector data to users;
- standards for spatial data; and
- spatial statistics and analysis.

Bridging the Gap Between Science, Policy, and Management

USGS mission statements include components that cross the boundary between nature and society, enhance and protect the quality of life, and contribute to wise development. To improve the nation's abilities to deliver a high quality of life and wise decision making, the committee urges the USGS to conduct supporting geographic research at the nature-society interface, in the Critical Zone. The Survey already provides valuable services to its partners and clients by supplying the spatial data they commonly use, but the appropriate role of the Survey in general and the Geography Discipline in particular includes fundamental research. The integrative power of geographic analysis and the communications power of geographic data could be substantially enhanced through research conducted nationally and internationally by the Geography Discipline.

Conclusion: The USGS is regionalizing its activities. This development positions the Survey to contribute to regional research and policy activities. To capitalize on this transformation the USGS should conduct substantive research that is explicitly regional, integrated, and place-based, in addition to its discipline-based research in geology, hydrology, and biology.

Recommendation: The USGS should strengthen its regional and place-based research (as opposed to topically divided investigations in geology, hydrology, and biology), including extensive involvement with regional research outside the Survey. The USGS should develop the ability to provide integrative regional experts for the nation.

Conclusion: The USGS cannot address all problems associated with bridging science, policy, and decision making, but its Geography Discipline can lead research activities in a few priority areas likely to draw upon existing expertise in the field of geography and improve the bridging function. GIS and remotely-sensed products promote citizen involvement at public meetings by providing a mode of communication between specialist and layperson based on data, while place-based frameworks and decision-support systems allow for experimentation to assist decision makers. Currently, the Survey lacks substantial research capability in these priority areas.

Recommendation: The USGS should assign high priority and substantial resources to fundamental research directed toward:

- improving citizen involvement in decision making for issues related to natural sciences by creating citizen-friendly geographic interfaces with all the Survey's primary spatial datasets;
- expanding the utility and application of place-based science by conducting integrative place-specific research in addition to topical research in individual disciplines; and
- enhancing the effectiveness of decision-support systems with increased geographic input and more effective map-like products as output.

SUMMARY

The USGS is reforming and incorporating missions that emphasize its role as one of the nation's most important natural science research agencies.

EXECUTIVE SUMMARY

The Geography Discipline produces valuable spatial data for users ranging from private citizens and corporations to governmental agencies at all levels. The Geography Discipline should now expand its activities to assume its proper role among the other disciplines at the USGS by engaging in fundamental geographic research, investigating the processes and forms that explain the dynamics of location, space, and place. The investment in such research will change the Geography Discipline, but it will pay enormous dividends for the nation by improving the science done in other disciplines, integrating new knowledge and data generated by the USGS and others, reducing losses from hazards, improving management of natural resources, enhancing the quality of life, and aiding in wise development. A strong Geography Discipline with a productive research component will ensure recognition of the USGS as scientifically credible, objective, and relevant to society's needs.

1

Introduction

Since the mid-1990s the U.S. Geological Survey (USGS) has redefined its role and organization in an attempt to improve its products and service to the nation. At the USGS, there has been an increased emphasis on the "Critical Zone," defined by the National Research Council (NRC) in its report *Basic Research Opportunities in the Earth Sciences* as the surface and near-surface portion of the earth system that sustains nearly all terrestrial life (NRC, 2001a, p. 35; Sidebar 1-1). This effort has included an attempt to become a revitalized component of a more inclusive natural science and information agency. By becoming a broader and more comprehensive organization, the USGS seeks the capacity to take a national leadership role by developing information and knowledge about the web of relationships that constitute the air-, water-, human-, and land-systems. To respond effectively to the challenge of providing objective, dependable information about the many science issues that affect human welfare, the USGS needs a strong Geography Discipline. In recognition of this need the USGS enlisted the National Research Council (NRC) to provide fresh perspective and guidance on future research and strategic directions for its geography program. This report distills the findings of the NRC Committee on Research Priorities in Geography at the USGS.

This introductory chapter begins with a brief description of the recent changes within the Survey. Subsequent sections in this chapter review the practice of science conducted at the Survey, outline the mission and vision statements that guide its recent reformation, and provide an overview of the report.

GEOGRAPHY AND THE CHANGING USGS

Established by Congress in 1879, the USGS was charged to describe the natural resources of the nation's western public lands. By the early twentieth century the Survey had become one of the nation's prominent natural science research agencies. Although geography's role at the Survey diminished shortly after the turn of the century, USGS researchers at that time were conducting other basic theoretical research and applied problem solving that was at the cutting edge of geology, geophysics, geochemistry, hydrology, geomorphology, cartography, and (later) remote sensing.

By the 1990s, however, critics in Congress considered the USGS obsolete and proposed its abolition. A thorough public debate about the Survey and its role resulted in a dramatic revision of those proposals. USGS clients stepped forward to convince their congressional representatives that the Survey continued to serve the public. As a result, Congress decided not to eliminate the Survey but instead to broaden its purview with additional responsibilities. These additional responsibilities came with the integration of the National Biological Service and parts of the U.S. Bureau of Mines into the USGS.

SIDEBAR 1-1
The Critical Zone: Earth's Surface and Near-Surface Environment

The surface and near-surface environment sustains nearly all terrestrial life. The rapidly expanding needs of society give special urgency to understanding the processes that operate within this Critical Zone (see figure below). Population growth and industrialization are putting pressure on the development and sustainability of natural resources such as soil, water, and energy. Human activities are increasing the inventory of toxins in the air, water, and land, and are driving changes in climate and the associated water cycle. An increasing portion of the population is at risk from landslides, flooding, coastal erosion, and other natural hazards.

The Critical Zone is a dynamic interface between the solid Earth and its fluid envelopes, governed by complex linkages and feedbacks among a vast range of physical, chemical, and biological processes. These processes can be organized into four main categories: (1) *tectonics* driven by energy in the mantle, which modifies the surface by magmatism, faulting, uplift, and subsidence; (2) *weathering* driven by the dynamics of the atmosphere and hydrosphere, which controls soil development, erosion, and the chemical mobilization of near-surface rocks; (3) *fluid transport* driven by pressure gradients, which shapes landscapes and redistributes materials; and (4) *biological activity*

INTRODUCTION

driven by the need for nutrients, which controls many aspects of the chemical cycling among soil, rock, air, and water.

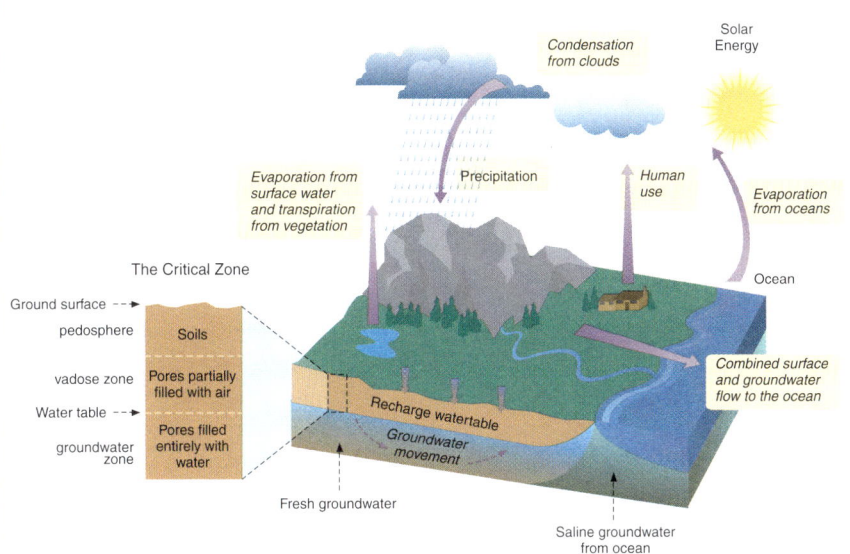

The Critical Zone includes the land surface and its canopy of vegetation, rivers, lakes, and shallow seas, and it extends through the pedosphere, unsaturated vadose zone, and saturated groundwater zone. Interactions at this interface between the solid earth and its fluid envelopes determine the availability of nearly every life-sustaining resource.

SOURCE: NRC, 2001a, p. 36-37.

This experience led the USGS, its scientists, leaders, and clients to re-examine the Survey's range of expertise, its niche in the federal constellation of research bureaus, and its broader societal role. The Survey directly serves the natural science research needs of the Department of Interior (DOI), whose numerous agencies work toward a broad array of mission goals and objectives; however, these agencies share a need for spatial data and natural science research. The USGS meets these needs through cooperative agreements with client agencies, producing geographic information-based products and by engaging in basic and applied natural science research on problems specified by the agencies. Examples include investigations into the physical effects of recreational use of public lands for the Bureau of Land Management, hydrologic analyses for water management decisions by the Bureau of Reclamation, and biologic studies of habitats and species for the Fish and Wildlife Service. The USGS also provides products for a variety of users outside the DOI who require mapping and other natural science data, including states, tribes, local governments, corporations, and private citizens.

The director of the USGS during the difficult years of the middle 1990s, Gordon P. Eaton, noted that the Survey persevered because of the value of its products to those who were willing to defend it. He also indicated that the inward-looking perspective of isolated researchers was no longer viable. "One lesson learned from the threat was that the viability and prosperity of the USGS depend on our ability to demonstrate the relevance of our work to society at large" (USGS, 1996). Eaton and his successor, Charles G. Groat, began a wide-ranging organizational and intellectual revision of the Survey to form a new social contract with the nation. In this new vision the USGS must respond to national and global needs by providing raw data, refined information, explanatory theories, and decision-support mechanisms to citizens and public agencies.

One outcome of this redefinition was the recognition by USGS leaders that geography again had a significant role to play in the Survey. When the USGS was first founded in the nineteenth century, geography was an integral part of its activities, but as the Survey evolved in the twentieth century away from pure description into basic earth science research, geography's role diminished. Internationally, European scholars were delving into spatial theory, but through the 1930s, U.S. geography remained largely descriptive. In the U.S. academic community geographers and geo-logists divided into their respective scholarly camps, and geography virtually disappeared from the Survey's activities.

By the end of the twentieth century, however, U.S. geography had academically become a powerful spatial science, and as a discipline it conducted scientific inquiry into the nature of space, place, location, patterns, distributions, and geographic flows. The discipline also included strong components related to the study of interactions between nature and society. The recognition of these potential contributions affected the reformation of the USGS. First, the newly defined Survey includes specific organizational accommodations for geography, including an Associate Director for Geography, a Chief Geographic Scientist, and a Geographic Information Officer. Second, the USGS now seeks to define and prioritize the potential bureau-wide contributions of geography in the Survey.

SCIENCE AT THE USGS

This report addresses issues related to the practice of science at the USGS. The term *science* denotes a way of knowing the world and a way of operating, a method of investigation with the ultimate goal of explanation and prediction. Science as a method begins with observations that may derive from field experience, theoretical manipulations, and data exploration. The collection and analysis of data then leads to classification and description, followed by questions, hypotheses, objective testing, and reporting of repeatable analyses. This series of steps in the scientific method is often so rigorous, and therefore laborious, that individuals or even organizations and disciplines may engage in only one or a few of the steps at a time. For biology, for example, more than two centuries was required to develop a descriptive classification of plants and animals before the science progressed to the next scientific step of intensive investigation and analysis. The USGS is charged to accomplish its basic mission of service to the nation through the practice of science, which requires a capability for sophisticated description, analysis, and dissemination of results.

In this report the term *natural science* denotes the broad range of scientific research addressed at the USGS. Natural sciences deal with the interactions, transformations, flows, distributions, and changes over time of both matter and energy. Decisions on priorities for the Survey within the natural sciences depend on the demands from other governmental agencies and society. A representative list recently published by the NRC characterizes the work of the USGS as including "geology, hydrology, geography, biology, and geospatial information sciences" (NRC, 2001b). The USGS often uses this same list in its vision statements. As the Survey becomes more involved in research addressing the nature-society interface, social processes will become more relevant to its work. In the present report the term *social sciences* refers to investigations of human activities, interactions, transformations, flows, and distributions, both individual and in social groups, that use the scientific method.

The reasons for specific scientific research at the USGS are based on external and internal factors. Types of studies and priorities addressed at the USGS are determined by (1) congressional intent, (2) institutional structure, (3) an agency focus on problem-solving, (4) magnitude of research projects, (5) expertise, and (6) data availability.

Congress often mandates certain research to be carried out by the USGS as part of the normal course of federal activities. This research can then be used to inform policy debate. For example, Congress has directed the Survey to conduct strategic assessments of oil and gas reserves. Alternatively, the research may provide the technical background for major public works, such as if Congress authorizes construction of large water management projects, with the planning and design work supported by USGS generation and analysis of hydrologic data.

Institutional structure is another determinant of the Survey's research domain. The USGS is part of the DOI, and its research first must support other DOI agencies (i.e., Bureau of Land Management, U.S. Bureau of Reclamation, National Park Service, and U.S. Fish and Wildlife Service that manage resources but do not have a research function. These resources range from minerals, oil, and gas to water for reclamation, parks for recreation, lands for grazing, and wildlife resources. A federal division of labor between research and management also results in institutional variation in responsibilities among agencies. For example, the National Atmospheric and Oceanographic Administration (NOAA) undertakes investigations related to atmospheric and marine processes. The National Park Service manages parks, but the USGS Biology Discipline conducts research on parks.

The USGS produces problem-solving research, with topics and subjects driven by problems posed to the Survey by its clients or by federal needs. This approach is different from some other research agencies such as the National Science Foundation (NSF), which funds many proposals for curiosity-driven research. The USGS also generates data, but such data are an early part of research at the Survey and are followed by intensive analysis, interpretation, and dissemination. This differs from agencies that are primarily or solely data producers.

Sometimes the USGS conducts research that is too large or complex for other entities to pursue. For example, geologic investigations into volcanic processes may require far more extensive instrumentation, travel, and maintenance in remote locations than universities or state agencies can provide, and projects that demand coordinated efforts of large numbers of researchers and support personnel are difficult for smaller institutions to justify. In many cases the USGS organizes, administers, and obtains funding for such efforts, with researchers from other organizations included in the effort.

In addition the USGS's research benefits from the Survey's personnel with special expertise. For example, the Survey's Biology Discipline includes researchers that form a critical mass of expertise in understanding the connections between endangered species and their habitats. Although individual researchers outside the USGS have similar expertise, the Survey has a large group of researchers with strong ties to national parks and national wildlife refuges. In addition, the USGS incorporates specialized technical expertise, including the nation's primary concentration of experts in stream gaging and earthquake monitoring.

INTRODUCTION

USGS research benefits from the Survey's long history of natural science data collection. The Survey's databases are national treasures that define the past and present characteristics of natural systems in the Critical Zone. USGS personnel know these data, understand their complexities, and can access them better than anyone else. Because effective science relies on observations and facts, the USGS databases provide a firm foundation for the Survey's investigations.

USGS VISION AND MISSION

Geography at the reformed USGS is part of a larger vision and mission for the Survey. The reformation of the USGS is a response to changing societal values that include more natural science requests in support of environmental quality, and devolution of federal authority with increased emphasis on service to regional and local clients. The vision and mission statements that stemmed from the Survey's self-examination reflect these new perspectives, which the agency explicitly states as follows (USGS, 2000).

Vision. USGS is a world leader in the natural sciences through our scientific excellence and responsiveness to society's needs.

Mission. The USGS serves the Nation by providing reliable scientific information to:

- describe and understand the Earth;
- minimize loss of life and property from natural disasters;
- manage water, biological, energy and mineral resources; and
- enhance and protect our quality of life.

These vision and mission statements give geography renewed relevance to the USGS's redefined purposes and show that the USGS has rediscovered interest in recognizing social relevance, in understanding nature-society interactions, and in using geospatial data, which are all major themes in modern geography.

Social relevance and the connection between society and environment are especially important. Geography and geology separated in the early twentieth century as geologists became increasingly focused on earth science. At a time when geographers studied Earth as the home of humans and sought to maintain their emphasis on general relationships between societies and their environments, geology became more specialized, and thus the two fields drifted apart both academically and at the USGS. The Survey seeks a closer association with geography to improve its approach to socially-relevant problems.

The refocusing of USGS interest on geospatial data is also a primary force in renewing geography at the Survey. Mapping has always been a primary activity at the USGS, but modern geospatial datasets go far beyond paper representations of spatial data. The ability to synthesize, display, manipulate, and analyze data directly from digital databases provide new scientific and decision-making tools that employ basic geographic principles. This combination of theory and methods results in a more complete product for the Survey's clients and partners.

With the new vision and mission initiatives in place, the USGS asked the NRC to undertake a comprehensive study of the evolution of the Survey and its future directions. In 1998 the NRC organized the Committee on Future Roles, Challenges, and Opportunities at the U.S. Geological Survey. The committee's recommendations helped sharpen the Survey's vision and mission concepts with specific statements on major responsibilities (NRC, 2001b):

1. The USGS should place more emphasis on multi-scale, multi-disciplinary, integrative projects that address priorities of national interest.
2. Information management at the USGS should shift from a more passive role of study and analysis to one that seeks to convey information actively in ways that are responsive to social, political, and economic needs.
3. The USGS should provide national leadership and coordination in (1) monitoring, reporting and forecasting critical phenomena, including seismicity, volcanic activity, streamflow, and ecological indicators; (2) assessing resources; and (3) providing geospatial information.
4. In addition to its high priority mission responsibilities, the USGS should shift toward the value-added activities of data analysis, problem solving, and information dissemination.
5. The USGS should continue to exercise national leadership in natural hazards research and risk communication.
6. The USGS should provide national leadership in the provision of natural resource information.

This suite of major responsibilities has implications for geography at the USGS in terms of the dominant geographic techniques of cartography, remote sensing, and geographic information systems (GIS). As the Survey becomes more deeply involved in acquiring, processing, and disseminating geospatial data, the skills of geographic specialists will become even more important. The design and production of maps evolves from a paper exercise to a cartographic design problem familiar to geographers. By suturing remote sensing to GIS, the Survey can apply geographic technical skills and provide products with raw data that have already graduated to the realm of information. The Survey also has the opportunity to contribute meaningful basic research in GIScience (GIS combined with spatial analysis) such as research to develop supporting theory for GIS, physical science for remote sensing, and mathematics for spatial statistics.

To implement the new vision and mission, Director Charles G. Groat charged a Survey Strategic Change Team to outline a managerial mechanism. The team's objective was to change the Survey from its loose confederation of divisions into a more closely connected community adhering to, accepting, and sharing the same vision and mission. The Strategic Change Team expanded upon the earlier general vision statement (USGS, 1999), stating:

> *The Strategic Change Team envisions a flexible and responsive USGS that is a recognized leader in providing natural sciences information, knowledge, and tools. Customers, partners, and USGS employees and managers will form an interactive community with a common passion to create, share, and use knowledge of the natural sciences to solve society's complex problems.*

To bring about such a vision the Strategic Change Team recommended a number of adjustments, the most important of which was to alter the organizational structure. Instead of maintaining a divisional perspective (e.g., geology, hydrology, mapping), the team proposed a Survey-centered framework view, which would emphasize multidisciplinary approaches and a regional structure rather than a topical one. Increased emphasis would be placed on management and decisions taken at the regional offices in Reston, Virginia (eastern); Denver, Colorado (central); and Menlo Park, California (western), thereby bringing the Survey closer to its clients and partners.

In response, the USGS eliminated the former division organizational structure (Biological Resources Division, National Mapping Division, Geologic Division, and Water Resources Division) in the fall of 2001 and replaced it with a discipline-based structure: Biology Discipline, Geology Discipline, Geography Discipline, and Water Discipline. Scientific efforts are under the direction of an associate director in each discipline at the national headquarters. The Survey includes associate directors for the Geology Discipline, Geography Discipline, Water Discipline, and Biology Discipline, and regional directors for each Discipline are at each regional office. The elimination of divisional boundaries in the Survey is intended to ease the formation of interdisciplinary teams and to improve the way complex problems that the Survey's clients and partners face regularly, are addressed.

Implementation of the reforms proposed by the Strategic Change Team and the development of disciplines have reopened opportunities for geographic contributions at the USGS in two ways. First, geography can provide the mechanism for integrating multidisciplinary studies and, second, the regionalization of the Survey taps the traditional strength of regional geography.

Geographic integration of data, explanations, and predictions is a logical approach for earth, water, and biological resources or hazards. Much of the data collected by governments at all levels has locational identifiers, so the association of geologic, hydrologic, biologic, social, and economic data is easier than ever before. The advent of highly sophisticated GISs, along with desktop versions useable by the non-specialist, further enhance the role of geography as an integrator for multidisciplinary approaches. Thus, geography and geographic data can provide a common language for specialist scientists working on common issues and problems.

Integration in the Survey and in geography elsewhere is often an exercise in regionalization. The regional organization of the USGS not only brings the Survey closer to its clients and partners, but also provides a source of regional specialists for its many clients. These individuals, reminiscent of the regional geographers who once integrated the natural and social sciences by geographic area, are potential facilitators for multidisciplinary investigations and management teams, as well as information contacts for the public. Regional geography can provide the Survey with a method of packaging data for multidisciplinary users outside the agency. The resurgence of regional interests in U.S. academic geography complements this trend toward regionalization within the USGS.

The relationship between geography inside and outside the USGS is a two-way connection. While geography plays a role in the new vision and mission of the USGS, the Survey can also affect the impact and progress of American geography. By providing the field of geography with an entry to federal natural science, the USGS can stimulate the best research, particularly in geospatial information science, natural hazards, and resource analysis. The USGS can become a critical contact point between the natural and social sciences, bringing these two perspectives closer in addressing the nation's needs. The Survey can connect its natural science data with social and economic data, especially from the Departments of Commerce and Transportation.

THE GEOGRAPHY DISCIPLINE AT THE USGS

The USGS was once divided into divisions related to subject matter, but during its recent reorganization the divisions were replaced with "disciplines." The Geography Discipline (once the National Mapping Division) has three programs: Cooperative Topographic Mapping, Land Remote Sensing, and Geographic Analysis and Monitoring. The Geography Discipline employs 1274 people, most of whom are technical grade specialists, and 369 contractors. The Geography Discipline's work force includes 10 Ph.D. geographers. Other disciplines may include a small number of additional geography Ph.D. holders. The number of Ph.D. researchers with degrees in subject matter coincidental with the Geography Discipline is small in comparison with other USGS disciplines. The Geology Discipline includes about 500 Ph.D. geoscientists; the Biology Discipline about 400 Ph.D. bioscientists; and the Water Discipline about 200 Ph.D. hydroscientists. The presence of these substantial numbers of Ph.D. researchers in the other disciplines provide those disciplines with a vital characteristic that the Geography Discipline lacks: a culture characterized by basic research ethics. These ethics include the skill to pose challenging questions, pursue rigorous analysis, and strive for predictions in a thoughtful agenda.

The budget for the Geography Discipline (a congressionally determined line item in the Survey's budget) has remained relatively level for the past decade, while other disciplines have increased until the most recent budget cycle. For FY2002 the budget was $133.3 million, approximately 15 percent of the total USGS budget. The Cooperative Topographic Mapping program received 61 percent of the Geography Discipline's total, Land Remote Sensing received 27 percent, and Geographic Analysis and Monitoring received 12 percent.

The combination of a history of emphasis on production, a lack of critical mass of Ph.D. researchers, and a budget that provides very minor support for its research and analysis program has created a Geography Discipline with a weak research element. This shortcoming occurs at a time when the Survey's vision and mission emphasize geography to a greater degree than at any time in the last century.

STUDY AND REPORT

This study and report constitute part of the reformation of the USGS. In 1999, while the Survey was seeking to refine its self-image, its disciplines of geology, hydrology, biology, and geography also began reviewing themselves and their futures. John Kelmelis, Chief Scientist for Geography, approached the NRC's Committee on Geography of the Board on Earth Sciences and Resources and suggested that the USGS could benefit from input by geographers outside the Survey. After the study was requested by the USGS, the NRC formed the Committee on Research Priorities in Geography at the USGS. The committee was invited to conduct a study on research opportunities in geography as they relate to the science goals and responsibilities of the USGS. It was asked to address the societal needs for geographic research and the appropriate federal research role. Specifically, the committee was charged to consider the following areas of concern to the Geography Discipline of the USGS:

- The role of the USGS in advancing the state of knowledge of the discipline (geography, cartography, and geographic information sciences);
- The role of the USGS in improving the understanding of the dynamic connections between the land surface and human interactions with it;
- The role of the USGS in maintaining and enhancing the tools and methods for conducting and applying geographic research; and
- The role of the USGS in bridging the gap between geographic science, policy making, and management.

In answering this charge the committee has created a report with four dominant themes:

1. The Geography Discipline should engage in scientific research.
2. The geographic research throughout the USGS should provide integrative science for investigations of the Critical Zone.
3. The Geography Discipline should develop partnerships within the Survey and with the field of geography outside the Survey.
4. Geography should develop a long-term core research agenda that includes several projects of the magnitude of the *National Map*.

As a foundation for geography's future contributions to knowledge detailed in subsequent chapters, Chapter 2 reviews the history of geography and geographers at the USGS. Chapter 3 outlines priorities for data and information management. Chapter 4 explores priorities for geographic information science (GIScience) as a means of maintaining and enhancing the geographic tools and technology of the Survey. Chapter 5 outlines priorities for research into the interactions between U.S. society and the land surface that supports it. Chapter 6 presents the committee's conclusions and recommendations.

2

Geography at the USGS

The future of geography at the USGS evolves against the backdrop of the past and present. Tradition is important as a precursor to change, and the present provides a starting point for the future. The review of geography at the USGS presented in this chapter is a primer for those unfamiliar with geography as a science or with the tradition of geography at the Survey. This chapter shows that present efforts to increase the contributions of geography at the Survey are not new; rather they are a return to a previously successful association between geography and the USGS. The chapter concludes with a brief explanation of modern geography in general, followed by a history of the field's association with the Survey and its personnel. The chapter continues with an overview of present research contributions in geography at the Survey.

MODERN GEOGRAPHY

Geography's Themes

Geography as an intellectual discipline is concerned with space, place, and location; and the distribution, arrangement, pattern, and flow of people and things in earth systems. One widely used definition is that geography is:

> ... *an integrative discipline that brings together the physical and human dimensions of the world in the study of people, places, and environments. Its subject matter is the Earth's surface and the processes that shape it, the relationships*

between people and environments, and the connections between people and places (Geography Education Standards Project, 1994).

Geography is not defined by subject matter alone. Geography shares an interest in physical earth processes with such disciplines as geology, oceanography, and meteorology, and an interest in human processes with such disciplines as sociology, anthropology, and economics. Geography is more clearly defined by its unique perspective on the world with its emphasis on spatial relevance. For the geographer, location matters most; other disciplines underscore the relevance of subject matter. Geology, for example, is the science of the earth, meteorology the science of the atmosphere, and sociology the science of human behavior and relationships. The definition of geography as a perspective on place is similar to the definition of history as a perspective on the world from the viewpoint of time. More than most disciplines geography views the world as a system with human and physical components operating in a complex set of interactions.

Because geography offers a spatial view of the world, much of the discipline's scientific effort goes to exploring how spatial processes operate and how spatial frameworks interact with other basic processes, such as those of physics, chemistry, biology, sociology, political science, and economics. Many geographers are scientists in the sense that they conduct their research according to commonly accepted methods of objective hypothesis testing using repeatable observations. However, some geographers investigate spatial interactions or nature-society connections using non-scientific approaches, employing techniques more closely aligned with the humanities. Because the objects of its study do not strictly define the discipline, natural scientists, social scientists, and humanists may hold the title of "geographer." The emphasis on spatial considerations has given rise to a particular set of tools and techniques for geographic inquiry, including cartography, remote sensing, and especially GIScience.

All phenomena exist in time and space. Therefore, they have a history and a geography. Consequently, the discipline of geography cuts across numerous scientific boundaries and disciplines. Yet, geography has a core and a coherence that stems from well-developed and widely shared approaches (NRC, 1997; Figure 2.1):

- A perspective that views the world through the lenses of place, space, and scale;
- Investigations and explanations that use three domains of synthesis: environmental processes, social dynamics, and a combination of the two; and

GEOGRAPHY AT THE USGS

- Spatial representation using visual, tactile, verbal, mathematical, digital, and cognitive methods.

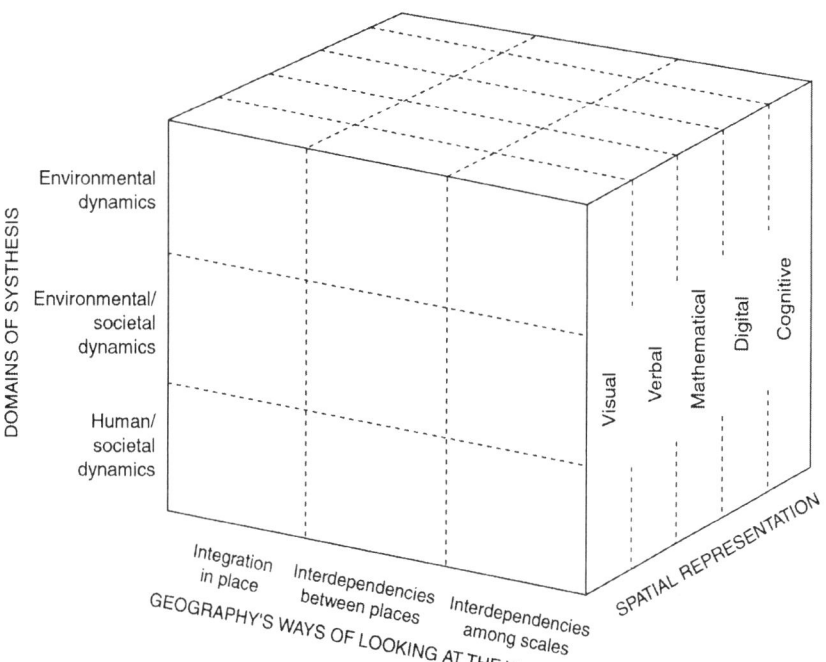

FIGURE 2.1 The matrix of geographic perspectives. Geography's ways of looking at the world—through its focus on place and scale (horizontal axis)—cuts across its three domains of synthesis: human-societal dynamics, environmental dynamics, and environmental-societal dynamics (vertical axis). Spatial representation, the third dimension of the matrix, underpins and sometimes drives research in other branches of geography. SOURCE: NRC, 1997.

These themes and approaches are important to this report, because they represent shared intellectual perspectives between the field of geography inside and outside the USGS. These concepts are also found in the vision and mission statements of the USGS, providing further connections between the Survey and the more general field of geography.

USGS Influences on U.S. Geography

Except perhaps in the domain of GIScience geography researchers outside the USGS do not interact much with the Survey, but this was not always the case. The central role many USGS geographers played in the early years of most geographic organizations in the United States illustrates the potential for future involvement in the larger sphere of geographic research outside the Survey. These experiences also show that there were strong, though largely forgotten, traditional geographic strengths in the USGS when geographers, geologists, hydrologists, and biologists worked closely with each other in resource surveys in the West (Figure 2.2).

Connections among the USGS, geologists, and geographers developed during the emergence of major professional organizations in the United States. In 1851, in an era of exploration and discovery, the American Geographical Society was formed to provide geographic information to business and government. The basic aim was to facilitate connections between the social world on the one hand and the natural world as represented by geology on the other.

Founded in 1888, the National Geographic Society became the best-known geographic organization to the U.S. and global public. John Wesley Powell (USGS Director), Henry Gannet (USGS Chief Geographer), and Clarence Dutton (USGS geologist) were all organizing officials of the NGS, and they were instrumental in defining National Geographic Society's simple but powerful statement of purpose, which still guides the organization: "For the increase and diffusion of geographic knowledge."

As geology and geography grew into more analytic sciences both fields required scientific societies for the exchange of formal scientific information, rather than the more general venues offered by the American Geographical Society or the National Geographic Society. Both fields were also dissatisfied with the general approach to science by the American Association for the Advancement of Science, where they shared an organizational section (Eckel, 1982). As a result, in 1888 a council of founding geologists, including Powell, established the Geological Society of America. At that time there were about 200 "geologists" in the country. The Association of American Geographers was founded in 1904 by geologic and geographic researchers who were also dissatisfied with the general approaches of other organizations.

USGS geologists played critical roles in the formation and continuation of the Association of American Geographers (James and Martin, 1978). The mainspring of the organization was its first president, William Morris Davis, who was a Harvard geologist adopting a more geographic perspective as his career evolved. Initially, its membership included numerous geologists, seven of whom served as president of the organization: Nevin M. Fenneman,

FIGURE 2.2 Geographers, geologists, and biologists of the Hayden Survey exploring the Yellowstone area of Wyoming in 1871. They exemplify the collaborative work among specialists in the western surveys that eventually were combined to form the U.S. Geological Survey. SOURCE: William Henry Jackson photograph 144, USGS Field Records and Photography Library, Denver.

G. K. Gilbert, Herbert Gregory, Francois E. Matthes, W. J. McGee, Rollin D. Salisbury, and Ralph S. Tarr. Although modern geographers might not recognize these names from the scientific literature, geologists would readily remember them as pioneering geologists, most with strong connections to the USGS. For the first several decades of its existence the Association saw geography primarily as the study of Earth as the home of humankind. Now in the early twenty-first century the vision and mission statements of the USGS place a similar emphasis on the perspective of the Survey. For geographers familiar with the Association, this "new direction" appears to be old (albeit valuable) wine in new bottles.

Throughout the middle decades of the twentieth century geology consistently progressed as a science, whereas geography continued its emphasis on description. By the 1950s geology had become an essentially quantitative science. At the same time, geographers had departed from the inclusion of the natural world in the systems they considered. Although regional descriptions inevitably began with the land, water, climate, and vegetation of a given region, there were few physical geographers who investigated these components of the earth system, and the division between geology and geography had become nearly complete. The USGS saw geographers as mappers—cartographers who created topographic maps—who provided little input into the scientific analysis of earth and hydrologic processes.

Beginning in the 1950s, however, the quantitative scientific revolution finally came to the fore in geography. Because the pool of physical geographers had decreased, this revolution most strongly affected the geographic investigators of social and economic phenomena, and the change was generally not visible to geologists. There were some exceptions: spatial analysis techniques were shared between physical and social researchers, but this was not widespread. In the ensuing decades the number of physical geographers increased, especially those engaged in geomorphology, climatology, and biogeography, and they adopted the scientific quantitative approaches now common to their areas of inquiry. At the beginning of the twenty-first century the geographer's research can be indistinguishable from that of geologists, meteorologists, and ecologists; the determining factor is the emphasis placed on the spatial component of explanation (the portion related to space, and location) or the nature-society connection. Researchers from the other disciplines often emphasize the aspatial aspects of processes, that is, the chemical, physical, or biological dimensions without respect to location. Geographers as a group are also more likely to include human activities in their explanations of natural processes than are researchers from other disciplines.

In the last two decades of the twentieth century an additional boost to geography came from the proliferation of GIS. GIS (remote sensing and mapping) and GIScience (GIS combined with spatial analysis) offered new and efficient ways to display and analyze geographic data. These powerful systems soon found their way into other fields as well, particularly geology, ecology, and climatology in the natural sciences, along with social and economic investigations. Resource analysts and planners began to use GIS to deal with urban, regional, and transportation problems, which further diffused geographic thinking, spatial perspectives, and an interest in nature-society connections by a variety of workers who had not previously dealt with geography. The USGS has become a strong player in the management and generation of the data used by such systems.

In summary the story of geography outside the USGS is significant because geography and the Survey were closely aligned but eventually drifted apart. As the USGS reunites geography with geology, hydrology, and biology in terms of research, instead of description alone, the Survey returns full circle to a previously productive arrangement.

HISTORY OF GEOGRAPHY AND GEOGRAPHERS AT THE USGS

Geography has not been highly visible at the USGS for more than 75 years. Consequently the rich history of cooperation and productivity enjoyed in the close association of geography with the other disciplines at the Survey early in its history is apt to be largely unknown to members of the Survey as well as academicians and professional geographers outside the Survey. A review of the story of geography in the Survey not only demonstrates the opportunities offered by a strong geographic component but also illustrates the ways geography can contribute to ensuring the social relevance of USGS programs.

The USGS originally was formed from four geographic and geologic surveys working in the regions of the U.S. West: the southwest; the interior basin and range, and Colorado plateau; the northwest; and along the fortieth parallel. After the Civil War these independent and distinct surveys began to inventory natural resources and to map publicly owned portions of the then largely unpopulated west. These regions comprised a natural resource base that would fuel the nation's economic development. Congress defined the survey areas and funded efforts directed by George Wheeler in the southwest, John Wesley Powell in the interior basin and range and Colorado plateau, Ferdinand Hayden in the northwest, and Clarence King along the fortieth parallel. In 1879 Congress ended competition for funding and influence among these individual surveys by combining them into the U.S. Geological Survey, under the direction of Clarence King. A year later John Wesley Powell became the director, and the USGS solidified into an entity that used both geology and geography as tools for the examination, classification, and assessment of the natural resources of public lands.

Geographic description in the form of topographic mapping was the focus of much of the USGS's early work (Figure 2.3). Because of the need for accurate locational data and exacting representations of the form of the land, topographic mapping was a prerequisite for geologic investigations. Topographic mapping was also the foundation for the Survey's hydrologic work, as knowledge about the patterns of the nation's watercourses was a necessary precursor to a useful stream-gaging program. ("Gaging" is the

FIGURE 2.3 Geography in the early USGS consisted of the creation of base maps for topography, hydrography, and geology, all founded on field surveys. This group of USGS topographic surveyors posed with their equipment during the survey of western public lands in the 1880s. SOURCE: USGS Field Records and Photography Library, Denver.

technical term used by hydrologists for "gauging" or measuring stream flow). The Survey's funding request for 1889 is an example of the distribution of annual funding during the USGS's first two decades. The request demonstrates geography's strong role by including about $200,000 for topographic surveys, $100,000 for geologic surveys, and about $100,000 for all other expenses combined (Rabbitt, 1980).

The importance of the early USGS's geographic or topographic work led to the creation of the position of Chief Geographer. Henry Gannett was appointed the first Chief Geographer in 1882, and he remained with the Survey in various geography-related positions until his death in 1914. Gannett's first task was the creation of a national topographic map that would serve a variety of purposes, including as the base of the national geologic map. He hired 78 cartographers and topographers (geographers), and geographers outnumbered geologists for nearly two decades (Rabbitt, 1980). In addition to his USGS work Gannett chaired the U.S. Board on

Geographic Names, was the Chief Geographer for the tenth, eleventh, and twelfth U.S. Censuses, and worked as assistant director and statistician for a census of the Philippines.

During the early years of the USGS, the difference between geographer and geologist was indistinct, and many practitioners considered themselves to be both. With a perspective that has re-emerged in recent vision and mission statements of the USGS, John Wesley Powell often styled his own career as one of geology and geography, because of his consuming interest in what would later be called the nature-society interface (Stegner, 1954). In addition he saw the early role of the USGS to include hydrologic hazard prediction, which he emphasized after the Johnstown flood of 1889.

Powell's most trusted associate, Grove Karl Gilbert, who held the position of Chief Geologist, also followed the dual model of geographer and geologist (Pyne, 1980). He projected the USGS into the disciplines of geography and geology outside the Survey. Today Gilbert is largely thought of as a geologist, yet he characterized himself as a geographer as well, and he was active in geography organizations. He was the only person to serve as president of both the Association of American Geographers and the Geological Society of America. The degree of his influence is now apparent; both societies recognize research excellence with "G. K. Gilbert Awards."

After about 1900 geography declined at the USGS. Powell had resigned as director in 1894, and Gilbert stayed with the USGS until his death in 1918. Gannett remained active, but he worked as much outside the Survey as he did inside. The USGS had changed its emphasis, focusing more on scientific research on geologic and hydrologic processes. By 1930 topographic maps (at a scale of 1:250,000) were available for most of the nation (Thompson, 1981), and attention along with an increasing percentage of the Survey's budget turned to using the maps as bases for geologic and hydrologic work (Figure 2.4). Although there was continued emphasis on stream gaging for water resource development and the geologic work that led to the development of minerals and fuels, there was a gradual shift away from other issues that were specifically related to societal concerns. During the first half of the twentieth century geology and hydrology grew into sophisticated sciences defined by their subject matter, while geography in the survey dwindled to descriptive cartographic work. Although mapping continued between 1920 and 1960, geographic research generally disappeared from the Survey.

By the early 1960s the moribund state of geography at the USGS began to change. In many respects during the previous several decades geography at the USGS reflected the reduced effectiveness of geography outside the survey. At the time academic geography began to progress rapidly into its scientific era, significant changes were occurring at the USGS. Owing to a

FIGURE 2.4 Creation of topographic maps in the early USGS required geographic technicians and much direct hand labor, as shown in this view of USGS map production facilities at the end of the nineteenth century. SOURCE: USGS Field Records and Photography Library, Denver, Colorado.

variety of new demands the discipline slowly began to return to visibility (Witmer, 2000). First, the U.S. Board on Geographic Names, which for several decades had been independent under the DOI, returned to the jurisdiction of the USGS. Second, the Survey began planning for the first national atlas, a project requiring a vast range of expertise and geographic integration. Third, researchers at the USGS became interested in remote sensing and in the quantitative spatial analysis used to exploit remotely sensed imagery from satellites (then a new technology). Fourth, the USGS made increasing use of classified geographic data that was created by senior geographers as part of their intelligence work during World War II. Fifth, demand increased from the Survey's clients and partners for remotely sensed geographic data on land use and land cover for mapping. Finally, the entire national perspective on natural sciences began to focus on environmental systems rather than on individual resource components. Geography theory and application were ready.

A reflection of this resurgence of geography at the USGS came with a return to the appointment of chief geographers, the first since the nineteenth century. As the second Chief Geographer, Arch Gerlach became the chief of the national atlas project and he directed the fledgling Geographic Applications Program, which was heavily involved in remote-sensing applications. With a much broader base at the USGS than the discipline had previously had, in the late 1960s geography consisted of two components: one largely descriptive and the other based on research and problem solving. In the 1970s James R. Anderson became the third Chief Geographer, and he oversaw a dramatic increase in the use of geography as an instrument for applied problem solving. Under his direction a national land use and land cover assessment program evolved, GIS and remote-sensing systems research flourished, and digital cartography emerged. Anderson increased the number of geographers working at the USGS, although cartographers and remote-sensing technicians dominated most of the new positions.

Anderson improved geography's ties between the Survey and the outside community. He was active with those in private industry and supported their emphasis on applied geography, meaning the application of geographic theory and technology to problems of social and economic interest. Anderson was active in the Association of American Geographers, and after his death in 1979 the Association created an "Anderson Medal" to recognize outstanding applied geographers.

Throughout the 1980s and 1990s geography slowly became more prominent at the USGS, although there was no Chief Geographer during much of that time. The Geography Program begun by Gerlach merged with the National Mapping Program and eventually became the National Mapping Division. The division was on an equal organizational footing with divisions for geology, hydrology, and later, biology, but its staff was smaller than the other three divisions. In 1995 the Survey's fourth Chief Geographer, Richard Witmer, undertook extensive efforts to modernize geography at the Survey, and he took up his predecessor's efforts to build ties with the geographic community outside the Survey. Links were established with state-based geographic information agencies and with the University Consortium on Geographic Information Science. He used two external studies, one by the National Research Council (NRC, 1997) and one by the National Academy of Public Administration (NAPA, 1998), to support the position that geography could make important, useful contributions to the Survey and the nation.

During this time geography at the USGS was also affected by an increased emphasis on spatial data and mapping requirements. Spatial data have become more common in the Survey and in the federal government overall, a trend that places greater demands on geographic analysis and data management. Digital mapping became a major effort jointly undertaken by the USGS and the Bureau of the Census, and the USGS became a major player in the new National Spatial Data Infrastructure (NSDI). The NSDI consists of the technologies, policies, and people necessary to promote sharing of geospatial data throughout all levels of government, the private and non-profit sectors, and the academic community. The initial description of the NSDI noted that it was to be the "means to assemble geographic information that describes the arrangement and attributes of features and phenomena on Earth. The infrastructure includes the materials, technology, and people necessary to acquire, process, store, and distribute such information to meet a wide variety of needs," (NRC, 1993).

In 1994 a presidential order directed that the Federal Geographic Data Committee (FGDC) control the NSDI and be housed in the USGS, with the mission of coordinating spatial data across all federal agencies. The FGDC builds partnerships with federal, state, and local governments to ensure a smooth flow of data in a standardized form that is readily accessible to a variety of users without unnecessary duplication (NRC, 1994). The USGS's expertise made it a natural leader with the virtual explosion in the use of GIS and remote sensing for the natural and social sciences and in decision making.

Another emerging geographic activity at the Survey is in mapping requirements. Previously most geographers were assigned to the task of completing the topographic mapping of the entire nation at scales of 1:62,500 or 1:24,000, a monumental effort that had begun more than a century before. As a reflection of the changing nature of geographic data, the USGS has also undertaken the task of creating an electronic national atlas, a digital version of the paper version. This national atlas is now available directly on the World Wide Web and provides users flexibility in creating individually tailored maps for specific applications.[1]

The emergence of spatial data and its analysis, along with increasing emphasis on nature-society relationships, play to two of geography's most enduring themes. However, as a result of the recent emphasis on cartography and mapping, the Survey cannot respond adequately to these research priorities. The Survey employs geographers in a variety of positions, mostly under job titles other than "geographer." Geographers at the USGS are mostly cartographers, rather than research scientists.

Personnel trained in geography are essential for developing the geographic information demanded by the Survey's clients and partners. Without

---

[1] The National Atlas (http://www.nationalatlas.gov) is a digital update of a large collection of paper maps that delivers easy-to-use map products to the public. It is distinct from *The National Map* discussed in Chapter 4.

additional experienced and highly skilled geographers in-house the USGS will be unable to provide comprehensive responses to societal needs. New hiring priorities should be based on the Survey's long-term core research agenda (NRC, 2001b), which has a significant geography component. The mission to "provide the nation with reliable, impartial information to describe and understand the Earth" (USGS, 1996) is similar to standard definitions of the general interests of geographers. Thus geographers who can provide leadership in the area of spatial analysis and on the integration of nature-society information should be recruited.

PRESENT GEOGRAPHY CONTRIBUTIONS TO THE USGS

Geographic research makes important contributions at the USGS (1) as an avenue of integration among the other disciplines; (2) as a form of research related to geographic data; and (3) in the creation of new tools for geospatial analysis. A summary of each of these roles reveals the present status of geography at the USGS.

Geography as an Integrator

Geographers are integrators of information and ideas from a variety of disciplines. Because of their integrative expertise, geographers often head multidisciplinary teams in attacking specific problems. A major method for bringing together disparate researchers and their activities is data integration, usually through one of three common approaches: historical, general systems, and geographic.

Historical approaches permit the organization of ideas, data, information, and ultimately explanations according to time. The logical arrangements of cause and effect emerge from time-based assessments of events, leading to attempts to predict the future by extending existing trends. Historical approaches can be worked in reverse as well, by observing present forms and processes and projecting them into past situations. Geologists rely on this approach with their motto of "the present is the key to the past," which has shown to be effective in unraveling Earth history by observing and measuring present processes. Hydrologists also search for historical trends in water resource data in attempts to predict future events.

General systems approaches rely on systems defined by elements and their functional interconnections. Analyses of elements and their connections as well as flows of mass and energy provide insight into the operations of complex systems, which are common in nature and society. In the natural sciences this is the ecological approach, wherein organisms and their inorganic support elements are viewed as elements and relationships within an operating system. For example, the fate of contaminants in a watershed can be described and understood by constructing a set of boxes (which represent the elements or compartments of the system) connected by arrows showing the rates of flow between them. The elements and flows can be quantified, with the objective of predicting the contaminant contents of each compartment.

Geography integrates data and information according to location, and it uses the map as the basic tool of representation. In a geographic approach to integration and synthesis each element and flow has geographic characteristics of location, place, arrangement, and direction of movement. The geographic perspective provides a significant tool for integrating data from a variety of sources and thus provides a common language for cooperation.

Geographic Data and Research

Geographic or location-specific data lie at the heart of much of the research conducted by the USGS. In science, description of phenomena is the starting point of the scientific method that ultimately results in explanation. Sciences mature as they graduate from an initial period consumed by description and classification into a period of more sophisticated hypothesis testing and analysis, followed by synthesis and theory building. For example, the science of ecology began with a period of "natural history." In this phase the classification of plant and animal species into elaborate descriptive systems occupied the time of most researchers. Later the science used those descriptions as a basis for insightful investigations of the environmental interactions of organisms and species. Geology followed a similar route, with early field descriptions of rocks being used to build spatial and temporal classification systems. By the late 1800s geology had become oriented more toward explanations that were scientific in nature.

Geography has gone through a similar evolution of scientific activity, but its embrace of scientific methods in the United States was much later than in other fields, such as ecology and geology. Topographic mapping in the 1800s was the geographic correlative to biologic and geologic description. Regional classifications were the geographic counterparts of the orders and families of species in biology or the terrains and eras in geology. Biology and geology progressed to quantitative scientific analysis at the turn of the twentieth century, whereas geography began that transition in the 1960s.

Geographic science is concerned with how geographic processes operate and how geographic patterns or frameworks influence human and natural phenomena. Examples of geographic science at work include: the spatial mechanics of diffusion; the influence of location and distance on the connections between entities; the topology of networks and regions; and how that topology influences physical, chemical, biological, and social processes. In nature-society interactions geography offers insights into hazards and resource perception as well as explanations for social outcomes of natural processes.

The result of the late emergence of geographic science is that many people outside the discipline are unaware of the recent advances in geographic theory and analysis. They continue to equate geography with pure description. If the USGS persists in viewing the potential contributions of geography and geographers as being solely in the areas of describing landscapes, stream networks, drainage basins, or hazards regions, the Survey will not benefit from the discipline's progress. Given that geography's basic proposition, that space, place, location, and the interaction between nature and society are keys to explanation, limiting the discipline to description will unnecessarily limit the Survey. In order to capitalize on the scientific advances the discipline has made outside of the Survey since 1960, the USGS will have to adopt new geographic research priorities to augment its descriptive capabilities in geography.

Geographic Techniques

When sciences emerge from their descriptive phases into activities more directly related to analysis, synthesis, and explanation, they often develop new tools and methods for measurement, display, and analysis. In the last 20 years geography has developed methods and tools of such utility and sophistication that they are widely used by the other natural and social sciences. The primary collection of geographic techniques and technology that support the modern revolution in geographic research include surveying, cartography, geographic information systems, remote sensing, and spatial statistical analysis. These tools form the basis of a geographic technology that is used in the acquisition, assimilation, manipulation, and representation of spatial data.

Surveyors, cartographers, GIS specialists, and remote-sensing professionals have ongoing concern with the acquisition, assimilation, manipulation, and representation of spatial data. The widespread adoption of digital technology and methods for management of large spatial datasets, has led to developments in all of these areas. Moreover, digital spatial information, which varies with scale, time, and spectral characteristics (the amount of electromagnetic energy exiting or emitted by an object or geographic area measured from a distance), presents unique theoretical and methodological opportunities for research.

The USGS conducts physical, biological, and social science research within its disciplines (Figure 2.5). Much of this science requires the collection of spatial information using *in situ* techniques, including personal observation and measurement with calibrated instruments. In addition, spatially distributed information is analyzed using technologies from cartography, surveying, and GISs. Dahlberg and Jensen (1986) suggest a model in which there is three-way interaction among the techniques where no sub-discipline dominates and all are recognized as having unique yet overlapping areas of knowledge and intellectual activity, the way they are used in physical, biological, and social science research (Figure 2.5).

Surveying

Surveying in the field for geologic, hydrologic, and biologic purposes has undergone a substantial revolution in recent decades. Although some traditional surveying is still accomplished using optical survey instruments, sight lines, and physical measurement, the Global Positioning System (GPS) is most commonly used now. With readily available hand-held instruments, this system commonly provides accuracy to within about a meter; increased accuracy is available with some additional instrumentation. The output from these field survey activities includes specific data personally recorded in the field along with precise location identifiers. For example, a researcher sampling vegetation types across a flood plain might use GPS to locate the ends of a 100-meter sample line and then count the number of individuals of a particular species encountered within three meters on either side of the line. Several such sample lines in a variety of nearby locations, each identified by the coordinates of its endpoints and with associated counts, become part of a dataset easily loaded into a GIS for mapping and further processing. All operations from initial sampling and counting to the final mapping and analysis are entirely digital.

Remote Sensing

Remote sensing is "the measurement or acquisition of information of some property of an object or phenomenon by a recording device that is not in physical or intimate contact with the object or phenomenon under study,"

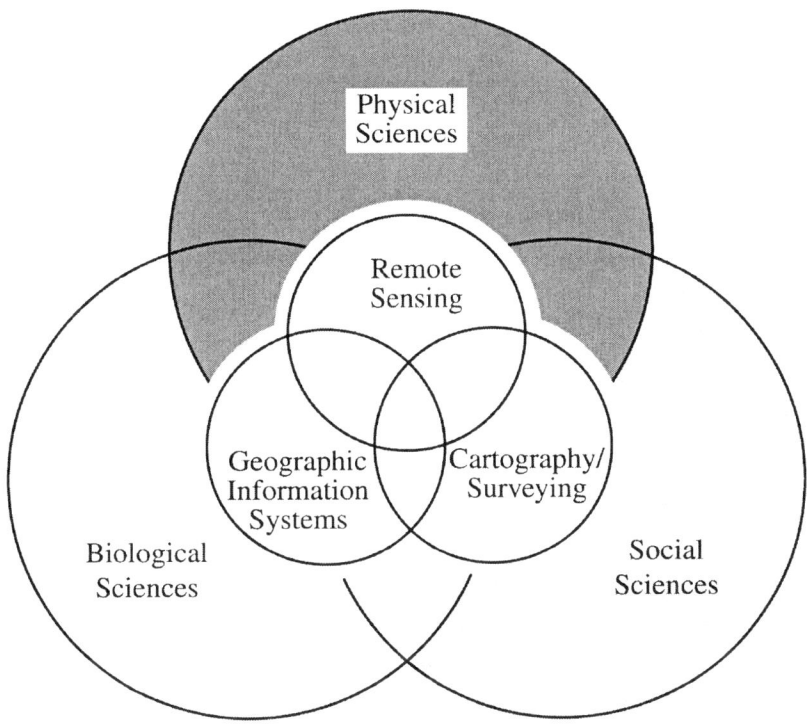

FIGURE 2.5 Venn diagram of physical, biological, and social sciences with remote sensing, GIS, and cartography. SOURCE: Jensen, 2000.

(Colwell, 1983). It is a scientific activity that uses sophisticated sensors to measure the amount of electromagnetic energy exiting an object or geographic area from a distance. These data are then analyzed to extract valuable information using heuristic logic and/or mathematical algorithms (Jensen, 2000). Remote sensing functions in harmony with other spatial data collection techniques or tools of the mapping sciences, including cartography, surveying, and GIScience (Curran, 1987).

The USGS has been involved in the use of remote sensing technology for many of its scientific endeavors since the 1920s. Scientists at the Survey have grasped the significance of remote sensing scientific technology and used it as one of their main methods of obtaining important cultural and biophysical information (Figure 2.6) and have made important contributions to science. Important ongoing USGS projects that involve remote sensing

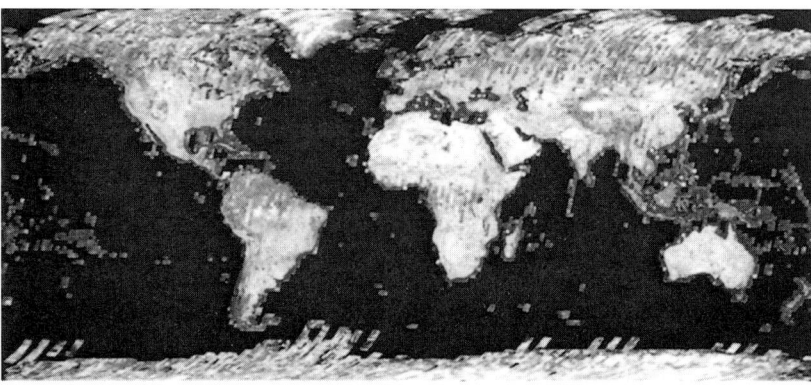

FIGURE 2.6 (a) The Landsat 7 Enhanced Thematic Mapper Plus (ETM+) multispectral remote-sensing system operated by the U.S. Geological Survey's EROS Data Center in Sioux Falls, South Dakota. (b) A mosaic of primarily cloud-free Landsat ETM+ images for much of Earth's surface. Individual orbital paths are discernible. SOURCE: USGS EROS Data Center.

include: (1) the development of an ambitious digital product, *The National Map*; (2) the development of the National Land Cover Dataset; (3) the creation of the global land cover dataset and the global forest map; and (4) major imagery archiving projects such as aerial photography, Landsat satellite data, Space Shuttle radar data, and accurate digital elevation models.

Cartography

Cartography includes the vital geographic functions of information representation and visualization within a spatial context. It underlies much of geographic analysis, and is concerned with all aspects of mapping. These aspects include: (1) the mathematics of projecting and abstracting (generalizing) Earth observations onto a map surface; (2) the cognitive and computational issues of geographic visualization; and (3) the design of algorithms controlling map production and dissemination (including Internet and related technologies). Cartographers also work to develop efficient ways of representing geographic features in digital format (Figure 2.7). They develop algorithms for efficiently storing complex digital spatial data, and for near-real-time data updates. Other USGS cartographic projects include hypermedia and animation methods for displaying patterns that change over time, the electronic National Atlas, and the World Construction Set (Figure 2.8).

GIScience

Geographic information scientists engaged in spatial analysis organize geographic information in a computer database, develop numeric models, and interpret spatial and temporal patterns in archived data (Longley et al., 2001). GIScience deals with the fundamental issues underlying spatial data and information technology. The software packages commonly referred to as GISs incorporate algorithms for the storage, retrieval, and analysis of geographic information. These algorithms are designed to handle large databases containing geographic data that vary in space and time. Spatial algorithms distinguish GISs from other types of information systems such as those that handle credit card transactions, personnel records, or other non-spatial applications.

FIGURE 2.7 A three-dimensional perspective of Kamchatka Peninsula, Russia, showing a cartographic representation of digital data. SOURCE: NASA Jet Propulsion Laboratory.

FIGURE 2.8 A digital cartographic image from the World Construction Set, showing diverse patterns over time. SOURCE: Dave Catts, USGS.

Spatial analysts build numeric models integrating biological, geological, hydrological, and social science data to study how location affects physical and human processes. They utilize spatial statistics to assess the reliability and validity of the models they create. For example, assessments of water supplies from alpine areas can use an image that presents a simplified model integrating terrain, rock type, and snow depth to predict availability of drinking water from snowpack in an alpine region (Figure 2.9). In other applications GIScience modelers overlay field data going back several hundred years to predict which vegetative types are most vulnerable to recurring severe forest fires. In a large-scale application the Gap Analysis Program (GAP) interrogates national datasets to determine and eliminate gaps in public lands and national parks so that habitats for endangered species can be preserved. GIScience in all its applications focuses on dependencies across space (spatial autocorrelation) and the influence of spatially distributed causal factors (spatial regression).

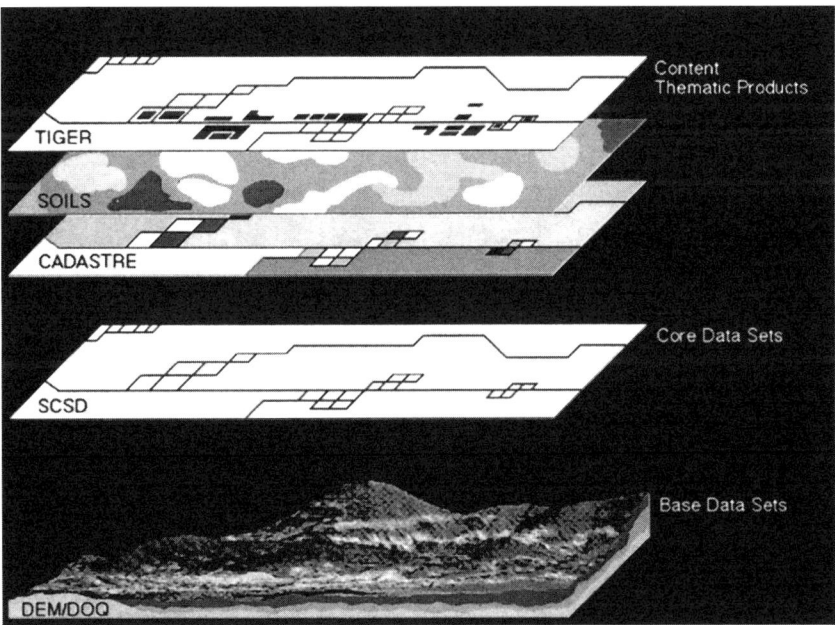

FIGURE 2.9 Digital data layers showing alpine terrain, hydrology, and vegetation, Niwot Ridge, Colorado. SOURCE: Larry Ayers, Intergraph Corp.

SUMMARY

Through much of the twentieth century geography at the USGS came to be synonymous with and virtually limited to mapping. Now the Survey faces challenges that result from three kinds of fundamental change. First, print maps have become much less important as end products for USGS while new digital products are increasingly important, resulting in a demand for an entirely new approach. Second, geography has changed as an academic discipline outside the Survey. Offering new perspectives on geography could contribute to the mission of the USGS. Third, the roles of the USGS have become more rich and complicated, as well as more strongly directed toward processes at the Earth's surface, necessitating a reassessment of priorities for the Geography Discipline.

From an operational standpoint geography is an integral part of the USGS's past and its present. During the early years of the Survey geologists and hydrologists worked closely with geographers. After a period of relatively little geographic activity at the Survey, geography is again a significant component of USGS activities. With the identification of the Geography Discipline at the USGS as an institutional partner with the Geology Discipline, Water Discipline, and Biology Discipline, geographic contributions in the future will likely be much more prominent. Selection of geographic research priorities will define the nature of geography's involvement in the reformed Survey. The following chapters explore these priorities in three general groupings: geographic data management, GIScience, and land surface-society interactions.

3

Priorities for Maintaining and Enhancing Spatial Data Management

For successful science, geographic research depends on specific tools and methods for acquiring, managing, processing, and analyzing spatial data. Since its inception the USGS has held responsibility for the basic geographic mission of mapping the surface and sub-surface features of the United States. Mapping methods have evolved from field-based surveys to include interpretations of hard-copy aerial photographs, and manipulation and analysis of remotely-sensed digital data. In response to changing customer needs the USGS's mission and vision statements articulate the need to accelerate the transition from traditional products (typically paper maps and reports) to new digital geospatial products. This chapter reviews the origin and types of existing geographic tools and methods at the Survey for data management and identifies priorities for their maintenance and enhancement. The following topics are primary and secondary priorities in spatial data management that coincide with the scientific mission and capabilities of the USGS:

Primary
- Collection and handling;
- Representation; and
- Integration.

Secondary
- Data mining;
- Historical data;
- Managing the security of data; and
- Spatial data reserves for decision making

48 RESEARCH OPPORTUNITIES IN GEOGRAPHY AT THE USGS

This chapter focuses on the data that support USGS research efforts in pursuit of its mission as one of the nation's leading producers of natural science information.

PRIMARY PRIORITIES

Spatial Data Collection

In addition to aerial and ground survey methods, imagery is the USGS's most common source of spatial data. The Survey began using aerial photographs for mapping in the 1930s (USGS, 2001a; Figure 3.1); by the 1970s satellite imagery became the primary source for capturing broad-scale geographic information (Figure 3.2).

FIGURE 3.1 Early aerial photography, a precursor to modern remote sensing, was an adventure. This pilot (W. Sidney Park, on the right) and photographer are about to begin a 1922 photography flight with the most modern photographic equipment then available, the small camera on the ground under the plane. SOURCE: USGS Field Records and Photography Library, Denver, Colorado.

FIGURE 3.2 Landsat 7 image of the United States. SOURCE: U.S. Geological Survey EROS Data Center.

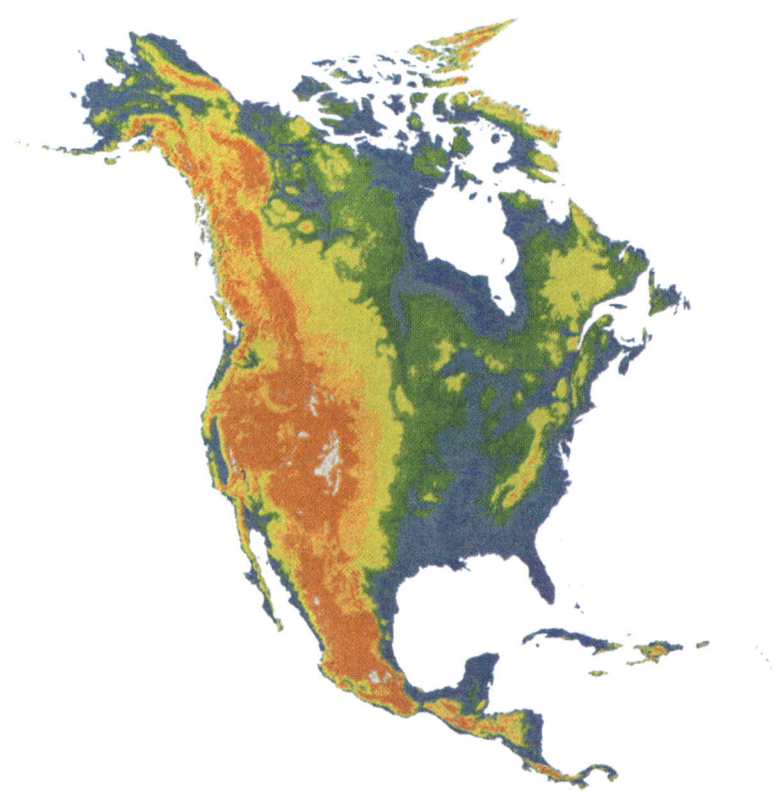

FIGURE 3.3 Digital Elevation Model, North American Continent. SOURCE: U.S. Geological Survey EROS Data Center.

The most important link between imagery and the final map is the Digital Elevation Model (DEM; Figure 3.3). Through the Geography Discipline (previously National Mapping Program) the USGS led the collection and dissemination of digital elevation data for the United States from the 1970s through the 1990s. In September 1999, the USGS completed national coverage (except Alaska) of 7.5-minute DEMs with data points at 30-m intervals. Toward the end of the mapping effort, most of the elevation models were produced by private firms, rather than the USGS.

As data sources change and improve, the trend in the USGS is to involve a greater number of participants in map and geographic information production and use. The National Digital Elevation Program (NDEP), for example, encourages partnerships with federal, state, local, and private sector organizations for building and maintaining digital elevation data. The USGS is positioned to provide the framework for geospatial information and should serve as the depository and portal for the Department of the Interior and other clients, and to furnish other information and derivative products in support of effective research and decision making (NRC, 2001b). One long-standing USGS data collection project is in Antarctica, where the Survey coordinates mapping and geodetic activities with the international scientific community under the terms of the Antarctic Treaty. The USGS maintains the Antarctica Geographic Names Data Base and the Atlas of Antarctic Research. Agency scientists, including geographers, collect Antarctic satellite imagery, GIS, gravity and glaciological data, and establish a geodetic control database for Antarctica (Draeger et al., 1997).

Successful data handling relies on standardization, especially when many users are involved. At the USGS, data transfer (i.e., the movement of data from one software or hardware system to another) follows spatial data transfer standard (SDTS) formats. In addition, the USGS is the lead agency for national elevation data, one of the seven framework layers of the NSDI. Finally, the Geography Discipline of the USGS produces a wide variety of digital spatial products for distribution, all using a common software platform.

Metadata are important components of spatial datasets that document the datasets and include the chronology of the data processing steps, extraction guidelines, desired and actual positional accuracy, and the temporal characteristics of the data. Metadata are an essential component of database management.

Spatial Data Representation

Digital data from aerial photography and satellites are represented as models, usually in the form of maps, for meaningful interpretation and archiving. The USGS uses two primary types of data models, raster and vector, for converting raw data into finished map and image products. In raster models the region is divided into rectangular blocks, or cells, with each cell containing a single measured value. Thus, a raster model for terrain elevation consists of a region of cells containing the land surface elevation within each cell. In vector models points, lines, and areas form the basis of representation for the originating data. In a vector model points are represented by coordinates, lines have starting and stopping points, and areas have lines as boundaries. Objects in a vector model therefore have characteristics or measurements that determine their appearance. An outline

FIGURE 3.4 Color-infrared color composite mosaic of the United States prepared from 16 images from the Advanced Very High Resolution Radiometer (AVHRR) sensors on the meteorological satellites NOAA-8 and NOAA-9. The images were acquired between May 24, 1984 and May 14, 1986. SOURCE: NOAA and USGS.

of states and counties of the United States, for example, can be represented in a vector model consisting of points and lines that describe the boundaries of each political division in the overall map.

At the present time the USGS provides paper and digital maps that subdivide the entire nation into topographic map quadrangles or special maps such as those defined by state boundaries. To examine data from large areas users may need to assemble several quadrangles or other maps; however, many of these maps do not fit together seamlessly. USGS researchers are designing new seamless databases that can be accessed without subdividing the information into topographic quadrangles. A seamless geographic database is more efficient to search, and researchers can download only the data they need instead of receiving all the data for a particular quadrangle. This sort of advancement in geospatial technology represents the next generation of maps, and the USGS should engage in fundamental research to support the approach in order to make it a reality. The seamless characteristic faces problems in matching features from one section of the map to adjacent sections, as well as connecting sections with different levels of detail.

Landscape models are now core research tools in the Biology Discipline and Water Discipline (Figure 3.4). Aquatic habitat patterns studied at the John Day Reservoir in Oregon combine a spatially explicit hydrological model with GIS data. Landscape models form the core of the Geography Discipline's urban dynamics research. Simulations based on algorithms derived from past patterns of growth can be used to predict future urban expansion. Links between these models and models of hydrological response or habitat quality could improve predictive capability. Urban growth models can also incorporate feedback from changes in the biophysical environ-ment. The committee urges the Geography Discipline to increase and streng-then its ties with the other disciplines in order to establish powerful integrative models with productive linkages. This objective is difficult because of the disparate vocabularies, scales, and methods of data collection used by various natural and social sciences.

Computational and content limitations in current GIS data models impede the generation of realistic landscape models of temporal geographic processes, such as stream discharge, chemical runoff through agricultural sites, and headwall erosion. Digital representations of geographic information lack appropriate data structures to handle elements with temporal variation in geometry or content. In addition, network models of spatial routing and capacity analysis are not integrated with visualization tools and mapping functions. For example, visualization tools could help transportation engineers monitor drift in iterative models and breakdowns in complex networks such as regional or national power grids, telecommunications networks, or sewage or water pipelines beneath city streets.

Emerging information technologies, such as animation and multimedia, introduce new requirements for cartographic design which remain underdeveloped and untested. Web interfaces could provide public access to data archives. Users would benefit from simple, easy-to-use interfaces for cartographic data made available on the Web. Without this access to public domain data for citizens and other users outside of the USGS, the nation's investment in improved data representation and analysis will not reach its full potential.

Integration of Spatial Data

In addition to collection and representation, another critical aspect of maintaining and enhancing spatial data is the integration of a variety of data types in a single product. For many years the production of 1:24,000 topographic maps was the framework for the nation's spatial data. The coverage of the continental United States by topographic maps was completed by 1991, but this paper-based method is now obsolete. It is impossible to adequately update the 55,000 topographic quadrangles in a timely fashion. Government agencies, companies, and private citizens require digital products. In response, USGS is developing *The National Map* (see Chapter 4). (The USGS refers to the programs and people in the Geography Discipline as "the National Map" but the committee believes it would be more appropriate to label only the product, "*The National Map*.")

The National Map demands a seamless database with thematic layers that will require substantial research and time for development. The first phase of the project is the creation of the National Elevation Dataset (NED), a seamless raster product for the entire nation that was assembled for the continental United States based on 7.5-minute Digital Elevation Model (DEM) source data (10-m and 30-m spacings). The committee recognizes that a major challenge is integrating source data other than standard USGS DEMs into the NED. For example, joining digital raster graphic (DRG) quads into seamless images is not a simple procedure. DRGs are reprojected to Universal Transverse Mercator (UTM) coordinates setting up a conflict between the data and the map frame. Attributes in a national seamless database must be rectified and standardized, creating another challenging research problem. For example, many resource classifications do not have clear, mutually exclusive definitions for each class of feature, and there is little standardization among the many state and private generators of such data. The effective management of these so-called fuzzy classes, construction of within- and between-class connections, exploration of relationships between classes, and evaluation of the results of these actions are formidable tasks. They require integrated database management, statistical graphics, and statistical analyses that are not yet fully developed (Fosnight, 1992).

SECONDARY PRIORITIES

Spatial Data Mining[1]

USGS clients require increasing amounts of data to solve research and management problems. Research on data mining can inform users about exploring and retrieving information from large data archives, but efficient and effective data mining remains an important unsolved problem (Ester et al.,1997). Many large databases currently being constructed contain spatial and temporal attributes, offering the possibility of discovering or confirming geographic ideas relevant to natural science and the USGS mission (Miller and Han, 2001). Decision makers can discover new data and improve the quality of their environmental policies by mining these existing archives (Yuan et al., 2001).

Relevant areas of data mining research include algorithms to browse the extensive archives of satellite data for predicting droughts, for evidence of pest infestations in forested areas, and for similar content-based pattern matching. Satellite information is being collected at such a rate that human inspection of individual data frames is no longer feasible. The USGS should develop automated data mining methods to provide timely natural science and hazards-related information.

Historical Spatial Data

The USGS has a long-established role of maintaining and archiving historical data. Its archives have become a valued resource for geographic research involving changes through time, and they are critical to numerous research applications ranging from hazard prediction and land use planning

---

[1] Data mining is defined as an information extraction activity whose goal is to discover hidden facts contained in databases. NASA Astronomical Data Center (http://adc.gsfc.nasa.gov/adc/adc_datamining.html).

to assessments of environmental change. Survey archiving and distribution of spatial data are guided by the Land Remote Sensing Policy Act of 1992 (15 U.S.C. § 5601 et seq.), and by rules established by the Executive Branch as the 1996 National Space Policy. Each policy requires the USGS to maintain a national archive (Draeger et al., 1997).

USGS archives contain aerial photographs used for topographic mapping with black and white photos dating back to the 1930s. The USGS's Earth Resources Observation Systems (EROS) Data Center stores negatives of more than 8 million aerial photographs from many federal agencies (USGS, 2001a). Also archived at USGS-EROS Data Center is the vertical color and color-infrared aerial photography generated by the National High Altitude Photography (NHAP) and National Aerial Photography Program (NAPP).

In addition to the aerial photography, USGS-EROS archives Landsat Multispectral Scanner (MSS) and Thematic Mapper (TM) imagery from NASA, Advanced Very High Resolution Radiometer (AVHRR) data from NOAA, and some declassified military imagery. The Center's holdings are expanding to include enhanced TM data from Landsat 7, digital terrain data from the recent Shuttle Radar Topography Mission, and data from new multispectral sensors aboard NASA's recently launched TERRA satellite. The USGS is now working with the SPOT Image Corporation to acquire and archive 10m-resolution images, primarily of North America (Kelmelis, 2001).

The USGS has generated most of the historical environmental data in the United States, ranging from stream discharge data collected by the Water Discipline to plant and animal data collected by the Biology Discipline. Methods of collection for land data varied from land survey records made a century and a half ago to the most recent satellite images. These data provide an indispensable context for present processes in the Critical Zone, the place of interaction between nature and society.

The federal government and others continually collect new information for monitoring the landscape. With the automation of data collection primarily through remote sensing, the rate of data collection has increased, with a concurrent increase in the amount of historical data. The rapid increase in historical data will be accelerated further with the advent of *The National Map*. All high-quality spatial data collected should be archived in such a way that it can be retrieved easily and used with modern GIS (Jensen et al., 1996). Metadata should also be archived and accessible. As technology continues to advance, storage media will change, data formats will be modified, and data from new domains will be collected. The committee urges that the USGS take a national leadership role in research on how to preserve the national geospatial data archives, how to manage access and update of this national treasure, and how to integrate new datasets and data domains as they become relevant. Furthermore, the committee notes that research is needed on how best to incorporate historical geographic data into the current digital formats.

Historical geospatial data include data collected or created in digital form (remotely-sensed data and digital cartographic datasets), as well as geospatial data in hardcopy (paper maps) that will need to be converted to digital form. The present demand and access needs are large [e.g., in FY2000, the Land Processes Distributed Active Archive Center (LP DAAC) distributed over 9,200 scenes (views of particular areas of the surface of the earth), consisting of 160,000 files or 1.1 terabytes of data to customers via FTP or the web] and this usage will continue to grow.

Historical spatial data priorities for the USGS lie in the following areas: digital conversions of hardcopy spatial data, research using historical data to assess human impacts on the natural environment, understanding slow environmental change, applications in ecosystem management and restoration, development of decision support systems, and finding solutions to methodological problems. Some of these priorities address research needs, others address data processing needs.

Digital Conversion of Hardcopy Spatial Data

Converting hardcopy spatial data into a digital format is a way to link the past and the future. The paper maps produced by USGS for over a century are a unique and important resource of topographic and land use data. The USGS can and should play a major role in the maintenance, digital conversion, and provision of U.S. historical environmental data from the pre-digital era. For example, in the restoration of the Everglades ecosystem of south Florida, USGS reconstructions of the original flows of water and plant communities provide a guide for engineering efforts to recreate a more natural hydrologic and biologic system in the region. As the Clean Water Act, Endangered Species Act, and other policies make environmental restoration an increaseingly common activity in the United States, there will be greater demands for digital reconstructions of former environments. One of the most important aspects of this task is digital conversion and georeferencing of the Survey's historical topographic maps. Digital con-version is the most efficient way to preserve these historic maps and make them accessible to users.

Other agencies will continue to archive their own datasets, but the USGS, through the FGDC, should provide leadership for data archiving, digital conversion, and access. The FGDC has established a Historical Data

Working Group, although it has not been active recently. The USGS should take a leadership role in this group and use it as an advisory body.

Documenting Human Impacts on the Environment

Historical spatial data play a significant role in the documentation of human impacts on the environment. These data can be linked with other numerical data to assess these impacts. For example, the USGS has historically collected data in tabular and text form on mineral production, stream flow, surface and ground water quality, water use, and other topics in the course of preparing early reports on resources and settlement potential. Other federal agencies that have been significant collectors of historical environmental data include the Bureau of Land Management (original land survey records for the central and western United States), the U.S. Forest Service (forest maps inventory from the 1930s), Agricultural Stabilization and Conservation Service (aerial photographs), U.S. Bureau of the Census, and the National Weather Service In addition, newspapers, company documents, and personal diaries and journals provide important historical data. Historical data have been used for environmental research for decades, but their use is now rapidly expanding.

Examples of human impacts that are traceable through historical spatial data are modification of land cover and vegetation, land drainage, wetland loss, changes in river channels, and filling along coastlines. Historical research often reveals unexpected, past land configurations or land uses that are not evident at present and are unknown to the present occupants.

Subjects for future research include topics that have received little attention in the past. For example, there is little research done in the United States on whether there are impacts from historical land cover change on water quality and quantity. Relatively little work has been done on the interplay of human-caused change and natural dynamics in geomorphic processes, such as coastal erosion and river channel change. Historical and ongoing land use change has major implications for the global carbon cycle, for energy and water transfer between the land surface and the atmosphere, and therefore, for climate change.

Understanding Infrequent or Slow Environmental Processes

Understanding of infrequent or slow environmental processes is enhanced by historical spatial data. For many environmental hazard processes, such as severe weather, floods, droughts, wildfire, and landslides, the primary source of information is direct instrumental monitoring or assessment of remnant geologic evidence. However, these processes typically have recurrence intervals close to or longer than the length of the instrumental record, and geologic evidence may be obscured by subsequent events. Therefore, historical records are critical for identifying and estimating recurrence intervals and possible magnitudes of the events. In addition, some environmental processes do not create immediate acute hazards, such as river channel change and shoreline retreat, and they may show relatively little change within the span of instrumental record. Historical time series are important for understanding processes that operate at rates not adequately captured by recent instrumental records. As long as data are appropriately archived, continued instrumented measurements will create a valuable historical record.

Historical records may also be useful in documenting human responses. For example, historical maps may be used to show the growth of flood protection measures (e.g., levees) that may influence the magnitude of subsequent floods. Historical data are important in assessing the character of nature-society interactions as well. Longitudinal data on the extent and value of built structures and other infrastructure are important for comparing the economic impact of different natural disasters in the same area. Historical data can contribute to the development of models displaying slow or infrequent environmental processes.

Ecosystem Management and Ecosystem Restoration

Access to historical data is often important for research on ecosystem management and restoration. Egan and Howell (2001, p.1) observed, "A fundamental aspect of ecosystem restoration is learning how to rediscover the past and bring it forward into the present – to determine what needs to be restored, why it was lost, and how best to make it live again." This perspective also applies to the broader challenge of bioregional assessment for ecosystem management.

A particularly important challenge in ecosystem management and ecosystem restoration is to determine the historical range of variation (HRV) for the ecosystem in question. Ecosystem status responds to natural disturbances that operate on time scales of years to centuries. The HRV represents the range of conditions and disturbance regimes under which the ecosystem is self-sustaining; if pushed beyond the HRV, the ecosystem will move into a state of disequilibrium, threatened species may be reduced or eliminated, and environmental quality may decline significantly. Historical spatial data are crucial to establishing the range of natural variation (Egan and Howell,

2001) and, in some cases, may be the only or best way of establishing a reference model for restoration or management.

Recently, historical studies have been initiated as part of regional ecosystem management planning. The Interior Columbia Basin Ecosystem Management Plan provides an instructive example: historical spatial data on forest management practices, including livestock grazing, timber harvest, mining, and road building from diverse sources provided an understanding of the causes of present forest conditions (Oliver et al., 1994). American Indian nations are involved increasingly in managing native lands, often with a goal of restoring to some extent the pre-European conditions and processes, and/or of reviving traditional ways of managing their landscapes (Egan and Howell, 2001, p. 14). Spatial data provide information on the extent and locations of traditional land uses, providing a guide to the process of cultural and environmental restoration.

Historical Data for Decision Makers

Often, the USGS utilizes historical spatial data to calibrate and test decision support systems. Historical data also are often used as input in simulations, which are critical to developing the scientific understandings that underlie predictive tools. Historical data explain how natural and social systems have responded to trends, stresses, and perturbations. In fact, analysis of historical data is essential for research on processes and behavior of social systems. The USGS's research on urban dynamics, based on historical data, illustrates the relative roles of natural and human-created features in shaping the spatial pattern of urban expansion. Such findings can enhance models of urban growth that may be used in decision support systems. Also of great value are the USGS's holdings of historical data on aquatic systems and wetlands, minerals production, energy resources, and other topics. The USGS should continue to work to make these historical data reserves useable in decision support systems.

Additional and primary methodological challenges in using historical spatial data are related to data availability and quality. Availability, level of detail, and accuracy decline as the age of data increases. The scale at which the original data were collected may not be the same scale as needed or as appropriate for the current problem, and data cannot necessarily be transferred across scales. Egan and Howell (2001, p. 13) argued that the best approach is to "combine appropriate techniques in a way that is multiscale and cross-referential in order to build convincing, corroborative lines of evidence." Because the USGS is the custodian for large amounts of historical spatial data, part of the Survey's mission encompasses basic research into a variety of questions that are historical in nature. In order to meet these responsibilities, the USGS should develop projects within the Survey and through cooperation with external researchers to address basic geographic research questions related to the accuracy, availability, quality, and scale issues for historical spatial data.

Managing the Security of Data

The policy of many government data centers, including the USGS-EROS Data Center, is to make their databases publicly available. However, in response to the terrorist attacks of 2001, federal agencies are reassessing public access to potentially sensitive data (DOI, 2001). Data that have previously been published may not be updated, and, in some cases, existing data have been withdrawn from the public domain. For example, National Imagery and Mapping Agency (NIMA) maps of certain areas in the United States are not released for general use.

The USGS also must be sensitive in the arena of geospatial data that might be used for targeting purposes. However, the restriction of general geographic data, including locational data and digital elevation models, can be detrimental to the basic mission of the USGS. From an economic perspective, the publicly available geospatial data provide important support for government management and private economic activity; sequestering too much data may result in significant disruption. The success of the American economy relies in part on available, low cost, public data. The archiving of valuable data, and the ready availability of those data for legitimate purposes, must not be abrogated unnecessarily. In many cases, it will be impossible to eliminate all public sources of geospatial information on the location of potential targets, so that the elimination of a single USGS source is not likely to result in improved security.

The USGS's general guidelines for assigning restrictions to data distribution (DOI, 2001) define a range of restrictions that might be assigned to documents and data (see Appendix D); these categories include data that are not restricted, partly restricted, or placed off-limits completely for public use. Decisions on restriction of distribution rest with the associate directors. While the current fundamental philosophy is to restrict distribution, distributing data is the driving philosophy of public agencies such as the USGS. In addition, since there are no substantive guidelines, the four associate directors' restriction categorizations have the potential to be inconsistent. The inherent conflict between security and the need for accessible information remains relatively unresolved because the policy articulated by the DOI is drawn from a legal perspective rather than a scientific one (DOI, 2001).

A uniform security policy for spatial data should be developed, and the associate directors should serve as advisors to a single USGS decision maker. To make as much data available as possible, the policy should clearly outline how the mission of the Survey and the security of the nation should be balanced in making decisions for data management.

Spatial Data Reserves for Decision Making

Good policy decisions that help maintain a robust economy and healthy environment are dependent on data that are as complete, accurate, and freely available as possible. Among U.S. government agencies, the USGS has been one of the major generators and providers of natural science and social science data and information for policy making. The USGS has extensive holdings of past data that will continue to be essential for contributing to informed policy making, and its new mission includes the continued collection of data vital for informed public policy. Before the advent of digital data collection and storage technologies, the USGS and other agencies had already collected a large amount of data, and these pre-digital are still extremely useful. Early topographic maps produced by the USGS include information on settlement, transportation infrastructure, river channels, and, to some extent, land cover and land use. In addition to first-generation topographic maps that may date from a century or more ago, revised editions of maps provide data at subsequent points in time that document changes from the first generation's measurements.

Knowledge about the topography, land use/ land cover, transportation network, hydrography, political boundaries, and place names of the American landscape over time is very important to a vast array of biogeographic, geologic, urban planning, and environmental spatial modeling applications. The integrated archive of this information has historically been the USGS topographic map series, which is currently archived in digital form as Digital Line Graphs (DLG) data in several series. These data form a record of changes to federally owned lands, terrain morphology, and transportation networks dating back over 100 years and provide a unique background for planning and decision making for all levels of government, private companies, and citizens.

Complementing these historical data will be an increasing quantity of new data. In the future, a considerable amount of spatial data is expected to emerge from federal agencies, state and local government agencies, commercial firms, and some private not-for-profit sources. A federal agency is the optimal steward for the archiving and preserving of these data for the nation. Although many federal functions have devolved to state and local authorities over the past two decades, management of the nation's spatial data infrastructure is a federal responsibility because placement in a federal-level agency is the only way to insure national uniformity in content, organization, and accessibility.

Although the provision of geospatial data was a government function in the past, the public and private sectors share this responsibility today. However, despite the increasing role of the private sector, local governments, and non-profit groups in geospatial data provision, the USGS should continue to be a major data provider in the future, in part because it is directed to do so by Congress as part of its mission, but also because it administers the nation's primary imagery and spatial data base archives. In addition, the USGS plays a critical role in developing standards through the FGDC.

SUMMARY

The USGS has a primary responsibility in the creation, archiving, distributing, and management of the nation's spatial data related to natural science. The role of the Survey in maintaining and enhancing these data includes the need for basic supporting geographic research, for special attention to historical spatial data and the development of security rules for their use and distribution, and for a continuation of the USGS leadership role with respect to spatial data and its standards. Data processing is not enough. Research into methods for data processing is the only way by which the USGS will improve its provision of natural science data and information that are relevant to society's ever-changing needs.

4

Priorities for Geographic Information Science

Chapter 3 explores the significant challenges and opportunities at the USGS for processing spatial data. This chapter addresses the maintenance and enhancement of analytical tools and the creation of specific geographic products. The chapter describes the Survey's primary priority as *The National Map* and defines the processing and research necessary for its creation. It also outlines the following more general secondary research priorities related to the generation and distribution of geographic products:

- resolution and scale;
- delivery of vector products to users;
- standards for GIS products; and
- spatial statistics and analysis.

These more general research priorities in GIScience are important because they present problems that must be resolved for two end uses. First, they support the completion of *The National Map*, which cannot become a reality until progress is made in these areas. Second, these general GIScience topics must be addressed if the USGS is to successfully produce an entire range of map products in new digital formats.

PRIMARY PRIORITIES

The National Map

Developing *The National Map* is the most important single initiative in the Geography Discipline at the USGS (USGS, 2001b). USGS administrators view *The National Map* as more than a single map or a static atlas. Rather, it will be a spatial database covering the United States and territories as a continually updated, "cooperative" topographic map. In this report the committee views *The National Map* as a digital product, a long-term research and production core project for the Geography Discipline. This definition is more restrictive than the use of the term by the USGS, where it is used to encompass almost all the activities of the Geography Discipline. The committee believes that the activities of the Geography Discipline should be more wide-ranging than *The National Map*, involving several lines of geographic investigation in the Critical Zone. Currently, the Geography Discipline compiles, integrates, and maintains databases that form the foundation of *The National Map*, but many additional datasets from outside the Survey will also be included. For example, street and highway locations and alignments will be derived from proprietary data, as well as federal data including human census data from the Bureau of the Census, agricultural data from the Department of Agriculture, and airport, railway, and port data from the Department of Transportation. The USGS is the appropriate agency to serve as the focal point of these various data streams and their expression in *The National Map* because the Survey is congressionally mandated to serve as the nation's manager of spatial data in the National Spatial Data Infrastructure. The USGS also has a historical foundation for its role as primary manager of spatial data. The Survey is the national purveyor of authoritative maps, and *The National Map* is a logical extension of that activity.

The proposed project is ambitious, requiring appropriate funding and considerable expertise in the processing of geographic information. The project is intended to provide integrated geospatial digital data for the nation, which will enable advancement in geographic research at the USGS, and promote the application of geographic information in decision making.

Unprecedented collaboration and partnerships among agencies, private organizations, and individuals will be necessary to develop and continually update *The National Map* (USGS, 2001b). Creation of *The National Map* involves much more than simply digitizing the current topographic maps; it requires a seamless geospatial database that has information from individual topographic maps restructured to achieve the goal of a map without edges. As discussed in Chapter 3, this information is currently available in fragmented form in individual topographic quadrangle maps, but the new database will require seamless integration of the information from the full set of quadrangle maps. Another goal of *The National Map* is the capability to update individual database elements as soon as change occurs in the landscape. The average age of the quadrangle maps is 23 years, and the current update cycles for paper topographic maps are unacceptable to modern users. Updating the paper-printed product is inefficient and costly. The goal for *The National Map* is to update weekly, essentially making the product a real-time spatial database. Given its other responsibilities for coordination and collaboration in map production, spatial data management, and the NSDI, the Geography Discipline should also assume responsibility for creating *The National Map*.

The database will have exceptional positional accuracy, with other federal agencies, private companies, and state and local entities providing information through the Cooperative Topographic Mapping Program (CTM). The inclusion of multiple partners means that the USGS's maintenance of data standards is a critical national function.

Although Chapter 3 discussed spatial data in general, the following subsections describe the specific data needs for *The National Map*. The USGS identifies the following data sets as the foundation for the map (USGS, 2001b):

- high-resolution digital orthorectified imagery;
- high-resolution surface elevation (topographic) digital data;
- vector feature data including hydrographic data, transportation data, structures, and boundaries.
- land cover data; and
- geographic place names.

The digital terrain information and the digital orthorectified imagery in *The National Map* are the foundation of spatial information in the NSDI (NRC, 1995). The digital orthoimagery is provided by the USGS, while the first-order geodetic control, from which digital terrain datasets are created, comes from the National Geodetic Survey. The National Geodetic Survey is a Department of Commerce agency within NOAA that provides accurate land survey and terrain elevation data by physical, on-the-ground methods. *The National Map* will grow from the NSDI, but will be a far-reaching extension of it. Major components of *The National Map* [for example, transportation, hydrology, cadastral (land ownership boundaries differentiating private from public parcels of terrain), and natural resources] are in harmony with the foundation and framework layers that reside in the NSDI (NRC, 1995). In addition, the committee identifies biogeographic data (the distribution of flora and fauna) as an essential dataset for *The National Map*. The committee did not include structures and boundaries in the list of priority research areas for *The National Map*, but these datasets are important to the Survey's clients and efforts to maintain and improve them should continue.

In the future *The National Map* will be more useful than paper topographic maps to the American public because it will be more accessible, more current, and will include a broader range of geographic information types from multiple sources. However, present knowledge, methods, and tools are inadequate to create *The National Map*. In order to achieve the project's goals, coordinated geographic research will be required in all three existing programs of the Geography Discipline: Cooperative Topographic Mapping (CTM), Land Remote Sensing (LRS), and Geographic Analysis and Monitoring (GAM). Additional research will require coordination between the Geography Discipline and the other disciplines at the USGS. The committee observes that there is no well-developed tradition of such interdisciplinary cooperation involving geography on large projects such as *The National Map*, but these intra-agency connections will be vital to a successful product.

The Cooperative Topographic Mapping program should have primary responsibility for the implementation and maintenance of *The National Map*, both challenging tasks that require research and development of new GIScience. Issues such as positional accuracy, seamless integration of disparate data from a variety of sources, resolution of scale issues, and construction of suitable database architecture require research prior to implementation. An additional challenge is the construction of an efficient Web-based interface for agency users and the general public. The Land Remote Sensing program will contribute Landsat and similar data to *The National Map*. LRS presently coordinates and distributes satellite and aerial photographic imagery for the nation. LRS's broader mission is to expand the use of these data. To do so, LRS must first coordinate remote-sensing data acquisition, processing, and distribution. These tasks have not been synchronized in the past, and federal efforts have been duplicated in independent projects that do not benefit from potential economies of scale. To coordinate these activities in a single program, LRS must also be able to advance remote sensing technology and improve the analysis of sensor records. Finally, LRS provides clients with better understanding of the applications and benefits of remote sensing. For example, the USGS's Earthshots Web site provides new users of remotely sensed products a basic introduction to applications of imagery for education, research, and problem solving (USGS, 2001c).

A focal point for project-specific geographic research at the USGS, the Geographic Analysis and Monitoring program (GAM) investigates the dynamics of environmental systems in the Critical Zone. GAM explores issues of sustainability, human health, natural hazards (including wildfire), and the geographic aspects of the global carbon budget. The focus of the program is on land surface dynamics. GAM applies the spatial, nature-society, and integrative perspectives of geography to USGS science.

Databases for *The National Map*

The committee identifies the following foundation data sets for *The National Map*: (1) orthorectified imagery (orthoimagery), (2) digital elevation data, (3) land cover data, (4) biogeographic data, (5) hydrographic data, (6) transportation feature data, and (7) geographic place names. Although the USGS deals with an enormous array of databases, these seven should receive the highest priority. Orthorectified imagery, high-resolution digital orthorectified imagery, aerial photography, and satellite imagery from which distortions have been removed should form the basis of *The National Map*. Such images combine the characteristics of a photograph with the geometric qualities of a map (Thrower and Jensen, 1976). Orthorectified images provide useful data on land use, built structures, vegetation patterns, and transportation features. Such images are often used by USGS and other agencies (for example, the U.S. Department of Agriculture for soil surveys) as a platform for the compilation and orientation of information from other datasets. Traditionally the USGS has produced orthorectified imagery of much of the United States from 1:80,000-scale National High Altitude Photography program imagery, 1:40,000-scale National Aerial Photography program imagery, or high-resolution imagery from satellites (Figure 4.1).

The use of orthorectified imagery in *The National Map* and in other digital map products of the future requires new knowledge. Orthorectified imagery research should include the development of methods to eliminate edge-match problems, and to unify the color balance throughout the dataset. Research should also be directed towards efforts to improve visibility problems in cloud-shrouded environments, perhaps by integrating orthoimagery produced from RADAR, or even by thermal infrared imagery. New methods of cartographic symbolization are needed that will enhance the visual utility of the images without obscuring the information shown on the base maps. High-quality metadata that describe the accuracy, processing chronology, and various sources of orthoimagery will be critical for analysis and sharing of orthoimagery.

Digital Elevation Data

High-resolution surface elevation data, including Digital Elevation Models (DEMs), are important by-products of the orthrectification process and provide the land surface for *The National Map*. The USGS has compiled terrain information for all of the nation's lands, but the coverage is at varying levels of detail. The most widely known map is the 7.5-minute quadrangle created from a DEM with a spatial resolution of 30m. For Alaska, most quadrangles are distributed at a scale of 1:63,360. Unfortunately, as the average age of the paper map series increases, much of the digital terrain data are out-of-date, particularly in areas subject to rapid change by human activities.

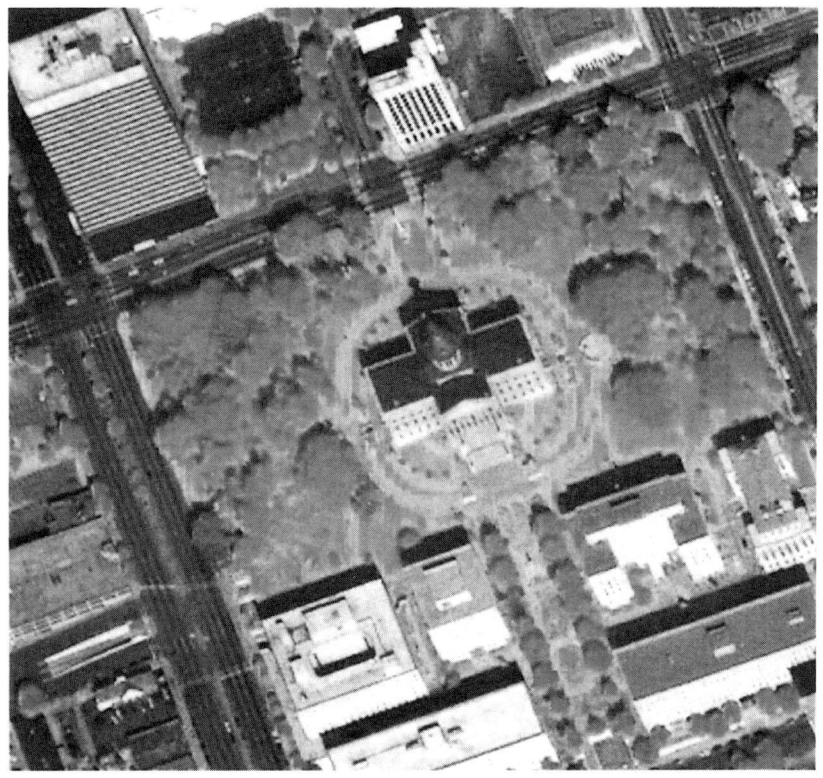

FIGURE 4.1 An example of an orthorectified image. A view of downtown Columbia, South Carolina, constructed from two separate images with different resolutions. IKONOS pan-sharpened 1x1 meter image courtesy of Space Imaging, Inc.

The USGS's current National Elevation Dataset (NED) is one of the most important datasets required for producing *The National Map*. The NED has grown out of the effort to create, from a variety of sources, a seamless global elevation dataset with a consistent geodetic datum at a nominal spatial resolution of 1km. The NED data are used for hydrologic modeling at the regional and continental scales, for transportation planning and development, and for geographic, geologic and biologic modeling problems such as wetland delineation, habitat management, and national response to environmental catastrophes (e.g., earthquakes). On average, approximately 20 percent of the 60-gigabyte database of 1-by-1 degree quadrangles is updated bi-monthly.

The initial compilation of the NED is complete (Figure 4.2). The USGS's researchers are now updating and refining the NED using a variety of federal, state, and local sources. Next, research should define the most efficient protocols for NED updates and refinements. For example, the USGS recently developed a derivative hydrological 1km dataset that includes slope and aspect information, which will be useful for modeling water flows and defining stream locations. However, this resolution is too coarse for identifying numerous smaller catchments, particularly those at high elevation. Limitations such as this can impede important national applications, such as predicting future sources of potable water for urban areas undergoing rapid growth. Improvement and refinement of a comprehensive national elevation dataset should be a high priority, and research is needed to guide this effort.

GIScience research is also needed to improve data integration algorithms for fusing the data from multiple agencies and for cleaning existing digital terrain data, which are prone to error accumulation during compilation. Management of data updating would also be improved with additional research. Increased emphasis should be placed on incorporating higher resolution elevation data from soft-copy photogrammetric techniques, light detection and ranging (LIDAR) remote sensing instruments, interferometric synthetic aperture radar (InSAR), and *in situ* point observations drawn from Global Positioning Systems (GPS) measurements.

Land Cover

General land cover information is required for many environmental, land management, and modeling applications (USGS, 2001d), and should be included in *The National Map*. Land cover data may be derived at a range of resolutions, depending on its source and its intended use. Spatially coarse land characterization datasets, such as those derived from the 1km Advanced Very High Resolution Radiometer (AVHRR), are well suited for global analyses. However, coarse-resolution data sets are often of limited value for regional investigations, and they are not appropriate for local land use studies. Although land cover datasets with very fine resolution (e.g., 1m by 1m cells or pixels) are appropriate for local land use planning, they are generally too voluminous for regional to global analyses (Vogelman et al., 2001).

In late 2000, the USGS EROS Data Center, in cooperation with the EPA, compiled the initial version of the National Land Cover Data (NLCD) set. The NLCD provides reasonably consistent and seamless 30m digital land cover data for the conterminous United States. It is an intermediate-scale national

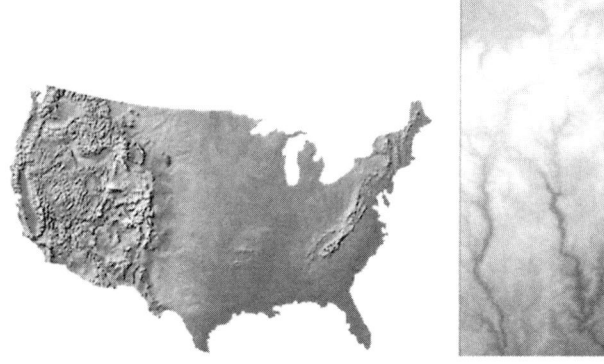

FIGURE 4.2 Mosaic of examples of the U.S. Geological Survey's National Elevation Dataset.
  A. A shaded-relief image
  B. Yosemite National Park in California at a spatial resolution of 2-arc seconds. The data have been rotated counter-clockwise 90 degrees.
  C. Yosemite National Park in California at a spatial resolution of 2-arc seconds displayed in an oblique, analytical hill-shading format with 2x vertical exaggeration.

land cover dataset for assessing water quality, ecosystem health, wildlife habitat, land cover, and land management issues (Figure 4.3). The NLCD was derived from 30m Landsat Thematic Mapper imagery and other ancillary data from the early 1990s, but significant research needs remain to be addressed.

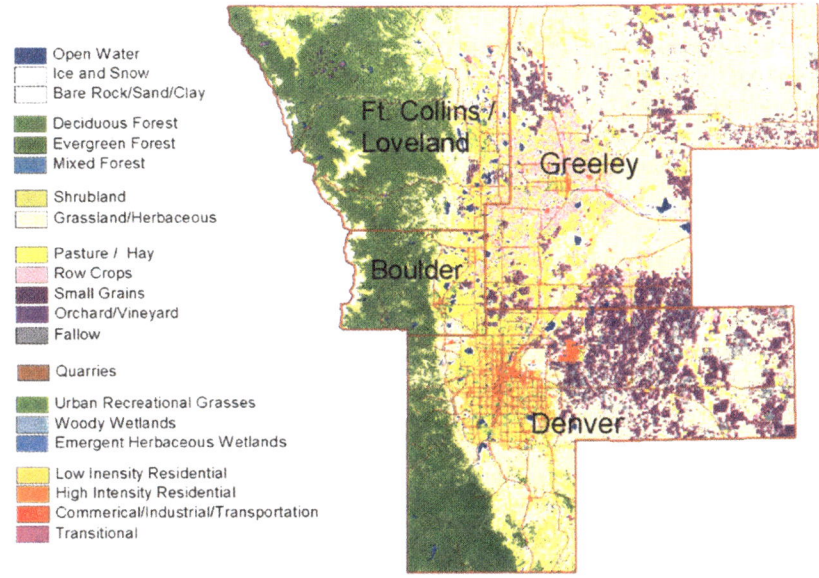

FIGURE 4.3 An example map created from data in the USGS National Elevation Dataset: a three-dimensional image of the Colorado Central Front Range near Denver, land use and land cover. SOURCE: USGS Front Range Infrastructure Resources Project; Data Source: USGS National Landcover Database (30 meter resolution), 1990.

Overall accuracy for the eastern United States was 81 percent for Anderson Level I aggregations (water, urban, barren land, forest, agricultural land, wetland, and rangeland) (Anderson et al., 1976), and 60 percent for all Anderson Level II classes (more finely divided classes, such as streams, lakes, and marshes for water features) (Vogelman et al., 2001). Since scientists and public agencies are generally unwilling to incorporate thematic data with only 60 percent accuracy, the committee believes that USGS geographers, in coordination with other scientists, should strive for 85 percent accuracy for Level II classes. This task represents a significant research challenge for the USGS. The USGS EROS Data Center has commenced refining the NLCD based on Landsat Enhanced Thematic Mapper Plus (ETM+) 30m data from early 2000. The Geography Discipline should use expert systems and machine algorithms to incorporate additional information in the NLCD, including canopy closure/density, soil taxonomy, permeability, and digital elevation data.

In the future, USGS geographers should integrate finer spatial-resolution remote sensing information into *The National Map*. Stereoscopic aerial photography with fine resolution will be needed to obtain accurate urban/ suburban land cover and land-use information. In this way, *The National Map* can achieve the accuracy and completeness required for modeling urban dynamics, land use change, and mitigation of agricultural and open space resources. Other types of supportive research required for land cover contributions to *The National Map* include improving change detection methods, developing error assessment mechanisms for the various classes of land cover, and making the data more accessible for general users by improving reporting.

Biogeographic Data

Data on flora and fauna along with their distributions should be included in *The National Map*. Ideally, databases for past patterns should be included and should be linked to models generating possible future biogeographies. The Geography Discipline should interact with the Biology Discipline and other organizations such as the Nature Conservancy to produce useful products for geographic science and the public good.

Biogeographic data from the Gap Analysis Program (GAP) should also be included in *The National Map*. GAP, coordinated by the USGS and carried out by state agencies, links predicted distributions of vertebrate species with natural land covers. The objective of GAP is to determine the effects of land management on the long-term maintenance of biodiversity. This program identifies species likely to be endangered by changing land uses. GAP data support and enhance the utility of other biogeographic data in *The National Map*.

Hydrographic Data

Hydrographic features are important components of *The National Map* for many applications in water quality, water use, land-use planning, and environmental management. The basic geographic building block for hydrologic systems in the Critical Zone is the watershed. The outlines of watersheds are major research, planning, and management tools for water and water-related resources such as habitat. Assessment of watershed resources, including water quality, has been part of the USGS mission since the Survey's founding, and support for watershed analysis has always been part of the geographic contributions of the Survey. In the late 1880s, the Survey developed a Hydrologic Survey to map dam sites and watersheds. Under the general guidance of the Water Resources Council, the USGS used several data sources to refine the basic geographic classification system for watersheds. The USGS created boundaries for 21 water resource regions, 222 planning regions, 352 accounting units, and 2,149 cataloging units (Seaber et al., 1987). These parcels of geographic space form the background for many potential applications using *The National Map*.

Watersheds are included in the National Hydrography Dataset (NHD), a geospatial vector dataset for the United States that includes all significant water features (USGS and EPA, 2000). Currently, NHD contains intermediate-scale data (1:100,000 scale) for most states and high-resolution data (1:24,000 scale) for some others. The hydrography of the nation is an important part of *The National Map* because many future users of the map will require this information.

Transportation Data

Transportation data describing railroads, airfields, highways, and city streets for *The National Map* are among the most valuable and widely used of all the data layers, but they are also among the most difficult to develop. These data are among the most needed components of the nation's data infrastructure. Linkage of geospatial transportation data to GPSs has stimulated automation and increased efficiency in emergency response, freight and mail delivery, and many other commercial routing applications. Highways and streets are not only conduits for the movement of people and goods, but they are also the alignments for a locational network of other data layers, particularly those for economic activity and population distribution and movement. The U.S. Bureau of the Census uses address files to supply locational identifiers for its fundamental data on human population, so that accurate locations related to street addresses represent the crossover point between transportation and socio-economic data of all kinds.

An important potential partner for the transportation layer of *The National Map* is the Department of Transportation's Bureau of Transportation Statistics (BTS), created in 1992. The USGS would be unlikely to contribute substantial amounts of transportation data to BTS, but is likely BTS would be a strong source of data for *The National Map*. The development of a USGS-DOT partnership requires improved interagency cooperation.

The maintenance of a transportation data layer creates one of the most serious challenges to *The National Map*'s objective of a near real-time database. The substance of many data layers, such as the land surface configuration, changes slowly over time, but transportation lines change rapidly. Within a single year, for example, many new streets are created, rural routes are abandoned, and significant changes occur in highway alignments as construction crews complete modifications. As the bridge network ages, load-bearing capabilities change and are reported at the state level by engineering inspectors. As a result of rapid changes such as high-way modification and load-bearing capacity, it is unlikely that the USGS will be able to update its own transportation database quickly enough to provide real-time data but will depend on acquisition of these data from other agencies or from private sources. For these reasons, the USGS should begin developing partnerships and promoting data standards for the transportation data needed in *The National Map*.

Geographic Place Names

The National Map will be useful only if accurately defined and located place names are included. The USGS is responsible for defining place names of natural and cultural features. Place names are a critical part of any geographic database because of their official sanction, which avoids confusing duplication with standard locations for each named feature. Each state has several-thousand place names, so that managing these data is a large task. The committee notes that *The National Map* should include the location of named features, and incorporate automatic and intelligent decisions about which names to include, and where to place them on the finished map. When users of *The National Map* zoom in or out, the system supporting the map should be able to automatically adjust its representation of place names to accommodate the user's needs.

Because of the connection between place names and the map, a gazetteer is an integral part of *The National Map*. A gazetteer converts the place name into geographic coordinates, permitting a GIS to identify a location on-screen, or to browse for information about the location in the data archive. Gazetteers preserve past as well as current names and support data retrieval for historical geographic research.

The Geographic Names Information System (GNIS), a digital gazetteer developed by the USGS in cooperation with the U.S. Board on Geographic Names (BGN), contains information on nearly 2 million physical and cultural geographic features in the United States. The federally recognized name of each feature described in the database is listed, and the feature's location by state, county, and geographic coordinates is defined. The GNIS is the nation's official repository of domestic geographic names information. The committee recognizes that additional research is required to translate data from GNIS into useful information for *The National Map*.

SECONDARY PRIORITIES

Resolution and Scale

Construction of *The National Map* and other digital map products requires a better understanding of how to deal with resolution of the original data and its final representation, because it will be a product that integrates data from a wide variety of sources. The content and geometry of vector data change with resolution. For example, tectonic processes are evident on maps at scales of 1:1,000,000 or coarser resolution that show vast areas of terrain simultaneously. More research is needed to fuse data from multiple sources to preserve geometric relationships (Quattrocchi and Goodchild, 1997). Research should also enable *The National Map* to model dynamic spatial processes.

Questions about the appropriate resolution also affect the creation and management of raster datasets. In some cases, the density of detail should vary within a single dataset. One example is the National Land Cover Dataset (NLCD), which must use the most appropriate resolution for each region of the nation rather than a uniform resolution for all regions regardless of complexity. For example, urban areas and strategic military sites might be compiled at a finer resolution than national forests. In particular, the question of how to develop detailed land-use information in the urban and suburban environment, and then include it in a more general database, remains a significant research need. It will be necessary to create automated techniques to extract information on structures (e.g., buildings, bridges), perimeter, and height from monoscopic and/or stereoscopic remotely sensed data. In addition, some of the information on urban infrastructure might more logically be derived from local LIDAR and InSAR data, both of which provide the exceptionally high resolution needed in complex urban areas. To make them manageable, the resulting datasets will require new algorithms for data compression.

Resolution is closely related to scale. The most obvious unsolved scale issues are connected to the need to use data from a variety of sources at different scales to create a product at a single scale. Additionally, all the data in any digital map are unlikely to be used at the same scale, so that when the user "zooms in" some previously unobtrusive data become apparent. The consequence of these scale-dependent problems is that it is impossible to create a single detailed national database containing all geographic themes (transportation, hydrography, elevation, vegetation, and so on) that would be needed to model every application at all possible levels of resolution. Because geographic data can be expensive to collect (e.g., data from extreme arctic or alpine locations) or available at unpredictable points in time (e.g., volcanic or seismic activity), data must be integrated from many sources at different levels of resolution. An important and challenging component of GIScience research is the integration or fusion of multi-scale datasets into a coherent and integrated database with established and usable scale limits for modeling, statistical analysis, and visualization. Multi-scale databases are often needed in a single application (Longley et al., 1999). For example, routing a national package delivery service requires geographic data at national, regional, and local levels to determine cost breakpoints for flying or driving the delivery routes.

Delivering Vector Data to Users

Much of the USGS's digital information can be viewed over the Internet using the Earth Explorer program. The USGS has been the world leader in developing systems for Internet access to digital environmental data, particularly through:

- Terraserver, providing access to the entire topographic map collection of the USGS;
- Earth Explorer, a gateway to the aerial photographic and satellite imagery collections of the survey;
- the National Atlas, a multiple-part mapping engine that allows users to customize individual map products; and
- the National Water Information System (NWIS), which provides geographically designed access to water quantity and quality information.

These systems have been extremely popular with users (they are accessed several thousand times each day), both inside and outside the federal government. Demand for digital data will undoubtedly increase as GIS technology spreads, and the quantity of data to be stored and accessed will grow exponentially with frequent revision of *The National Map*.

Transmitting geospatial data via the Internet has become commonplace, usually using raster data that is efficient for imagery, orthophotography, and satellite data products. Transmission proceeds incrementally: transmitting additional raster rows and columns in random order refines an initial coarse resolution. The computations are straightforward, and file size remains predictable and constant for a given resolution, regardless of feature complexity. At the viewer's workstation, the Internet browser displays the visual effect of a blurred image whose details gradually sharpen.

Unlike raster data, vector data remains a challenge for Internet transmission (NRC, 1999a; Buttenfield, 1999). Vector files tend to be large, and file size increases unpredictably depending on the complexity of feature geometry. It would be ideal to transmit vector data by first sending a coarse "browse version" and then amplifying details. This problem is a high priority research area because 911 emergency response systems and other uses require real-time access to vector data. The USGS should pursue research to improve transmission of vector files because they are essential for most of the digital products likely to be produced by the Survey.

Standards for GIS Products

Standards for geospatial products address data lineage, accuracy, the compilation sources, the processing methods, and the chronology of processing, as well as error and uncertainty. The elements of standards for geospatial data are positional and attribute accuracy, completeness, logical consistency, and lineage. The description of these elements is in the metadata for any geospatial product. Initially, the FGDC developed data standards for U.S. agencies to share national geographic information. The creation of the standards was a monumental organizational achievement involving government, private sector, and academic participation. The resulting Spatial Data Transfer Standard (SDTS) was incorporated as Federal Information Processing Standard (FIPS-173) by legal mandate (NIST, 1994). The NRC has been continually involved in the standards process through its Mapping Science Committee.

Assessing the reliability of geographical data for inference and reasoning is an important and emerging area of research in GIScience. With development of GIS and related technologies, large volumes of data may be displayed and analyzed rapidly. But interpretive capabilities depend, to a certain extent, on the ability to visualize accurately and to discern the quality of the identified patterns. Data quality influences the credibility of data analysis and the confidence attached to data interpretation. It affects the reliability of interpretations and thus decision-making based on GIS modeling, sensitivity analysis, and data exploration. Data quality usually varies across the map surface, and this variation needs to be communicated to the map user. Data producers, such as the USGS, have a responsibility to advise data users about the quality of generated data products. USGS data products form a foundation for many critical decisions and policies about the nation and the environment. For example, data collected in ecological field studies are difficult to ascertain as being positionally accurate, correctly categorized, complete, or logically consistent (Buttenfield, 2001). Nonetheless, these observations affect policies on wetland delineation, habitat loss of indicator species, and similar issues affecting urban corridor development and spread. The USGS should conduct research on how to model spatial variation in error, how error is introduced into geographic information during data processing, and under what circumstances error may be reduced by particular computations.

Research underlying the development of data standards includes many components of data production that at first seem unimportant. For example, standards include rules for thesaurus development to ensure the consistency of data definitions (geographic meaning as well as semantics) (Salge, 1995). Thesaurus development becomes especially important when a dataset is used for multiple purposes. For example, the definition of an "address" to the U.S. Postal Service is the coordinates of a mailbox. The definition of an "address" to the 911 Emergency Dispatch teams across the nation is the coordinates of a front door. In urban areas, the geographic distance between a front door and a mailbox can be measured in inches. In rural areas, the distance may extend a quarter mile, or more. When the Postal Service and 911 Dispatch first exchanged national datasets, both agencies intended to reduce their cost of digital data compilation. However, the differences in data definitions of an "address" made it impossible for the agencies to distinguish semantic differences from database errors (Buttenfield, 1997).

Research on GIS standards has progressed in other countries. With international collaboration for creating global databases, including in the International Geosphere-Biosphere (IGBP) Project, the need for data sharing with other national governments complicates the standards development procedure (Wortman and Buttenfield, 1994). Specifications for data production may differ dramatically. The differences in positional accuracy for cadastral data compiled in East Germany and West Germany before the fall of the Berlin Wall were not greater than 16 centimeters. Yet when the two national databases were integrated, the positional discrepancy was propagated across hundreds of kilometers, and entire land parcels dropped out of virtual existence in the new database. This problem had important implications for the landowners and for the newly unified German government as it tried to establish demographic inventories for social services and taxation.

Priorities in Spatial Analysis

The successful creation and use of USGS digital products for geographic data require a sophisticated capability in spatial statistics and analysis in two ways. First, the USGS personnel must know about analytic techniques if they are to create products that lend themselves effectively to analysis. Second, USGS personnel must possess substantial analytical capability to address the Survey's vision and mission, both heavily reliant on spatial data.

Spatial analysis uses transformations and statistical manipulations to identify trends, reveal patterns, and detect outliers or extreme values. Spatial analysis also provides computational guidance, confirming the presence of a pattern and its significance in situations where visual displays may confuse or deceive the researcher. The distinction between spatial analysis and conventional statistical analysis is that the former assumes that the results of a method will vary as the location of the objects under scrutiny varies, and location can be defined in terms of location on the planet or in the database. Longley et al. (2001) identified six areas of spatial analysis: queries and spatial reasoning, geospatial measurements, data transformations, descriptive summaries, spatial optimization, and hypothesis testing for spatial pattern. These areas, which use spatial inference, are enhanced by and rely on effective research on the techniques of spatial analysis.

To support the Geography Discipline at the USGS, fundamental research in spatial analysis is needed. Research in spatial analysis will improve understanding of spatial scale, spatial association, spatial variation, and spatial movement. Such research will also improve the Survey's service to users by improving query and retrieval processes as well as the handling of very large spatial datasets (e.g., disaggregated census data and remotely-sensed data on a global scale). Likewise, appropriate tools to model geo-graphic processes will be developed through such research. In the United States, enormous quantities of data are now available to help solve local and regional problems, but the magnitude of the data sometimes makes them difficult to use. Research on spatial analysis will help researchers and decision-makers use available data optimally, even when those data are in terabyte- or petabyte-sized collections (UCGIS, 1998). As satellite imagery archives, digital government repositories, and other georeferenced datasets become larger, methods should be developed for browsing, classifying, and manipulating the information efficiently. Real-time database access can improve support for emergency teams such as national 911 Emergency Dispatch services designed to speed responses to natural, technological, or terrorist hazards.

To help achieve the vision and mission of the USGS, the Survey should improve its contributions to geographic knowledge, tools, and techniques by developing the capability to address the high priority subjects of resolution and scale, delivering vector data to users, standards, and spatial statistics and analysis.

SUMMARY

This chapter explores the role of the USGS in contributing to new knowledge in the areas of cartography and GIScience, and delineates the priority research items. *The National Map* is expected to be the flagship product of the Geography Discipline in the next decade. However the accomplishment of this worthy goal is not assured with present knowledge. Further basic GIScience research is needed to make the product a reality. The USGS should undertake additional GIScience research towards the development of *The National Map* and of a more general nature, if it is to fulfill its vision and mission.

5

Research on Land-Surface and Society Interactions

The USGS vision and mission statements point the agency toward an expanded role in data management (Chapter 3) and GIScience (Chapter 4). An additional dimension of more general research for the USGS's Geography Discipline emerges from the agency's vision and mission statements. The following topics are primary and secondary priority research opportunities for the USGS:

Primary
- Environmental resources and systems; and
- Natural, technological, and security hazards.

Secondary
- Urban dynamics;
- Regional and place-based research; and
- Bridging science, policy, and decision making.

This chapter describes the nature of USGS contributions to each of these research lines and explores potential activities that could contribute new knowledge and build bridges among science, policy, and decision making.

The interplay between human and natural systems, the subject of this chapter, occurs most commonly at and near the earth's surface, a portion of the earth environment characterized as the Critical Zone (NRC, 2001a; Sidebar 1-1). Processes in the Critical Zone include human population growth, increasing demands on physical and biological resources, and declining health of environmental systems. The natural systems also pose risks for human society in the form of landslides, floods, coastal erosion, and a host of other hazards. The Critical Zone is the dynamic interface between the solid earth, the hydrosphere, and the atmosphere, connected by a complex web of linkages, feedbacks, reservoirs of material and energy, and chemical interactions. The USGS should lead in contributing to basic scientific understanding of the Critical Zone, especially in creating ways to employ geography's integrative capabilities. This role is appropriate for a federal agency engaged in natural science information and research.

The committee assessed the geographic initiatives undertaken by other agencies, and it considers the research agenda for geography presented in this report appropriate for a federal natural science agency and complements the work undertaken by other federal natural science agencies. Geographic research at the USGS is important as a service to the nation because the agency is unique in its experience and resources. Geographic issues in the public arena, ranging from the management of public lands and waters to assessment of hazards rely on geospatial data that are largely the product of the USGS. The USGS is the only federal agency that has both (1) a mandate to provide the nation with natural science information and knowledge and (2) a large cadre of specialists in geography, geology, hydrology, and biology.

Geographic research is administratively housed in a number of organizational units at the USGS. Although some geographic research activities are undertaken within the Geography Discipline, other geographic research is conducted within the Biology Discipline (e.g., research into climatic change or scale and pattern in ecosystems), Geology Discipline (e.g., research in epidemiology), and the Water Discipline (e.g., research into river systems). The initiation and conduct of geographic research in the USGS is important in addition to its organizational placement, as geographic researchers is distributed throughout the agency.

PRIMARY PRIORITIES

Environmental Resources and Systems

Basic and applied research on environmental dynamics has long been a major focus of physical geographers in universities and research institutes (NRC, 1997). Physical geography has most often been concerned with geomorphologic, hydrologic, climatic, biogeographic, and pedologic systems. Research on how natural systems operate, the interactions among them, spatial patterns of forms and processes, and the human influence on these systems are classical themes in physical geography. A major theme in geography has been the study of nature-society interactions. Study of the interactions among these human and natural systems is an emerging opportunity in the more general Earth sciences, because of the importance of human effects on environmental systems and vice versa.

Three general research issues emphasizing the spatial aspects of environmental systems, or the interaction of society with those systems, are exemplified by:

- Climate change;
- Scale and pattern in ecosystems; and
- Integrated studies of rivers systems.

Climate Change

Over the next 50 to 100 years climate change will affect many if not all environmental and human systems. Possible manifestations of climate change include not only shifts in climate regime (temperature and precipitation) but also increased climate variability and increased frequency of extreme events such as hurricanes, droughts, and severe winter storms. Using the past as key to the future is a fundamental tenet of climate change research. Understanding past climate change on a decade-to-century time scale is an important component of the interagency U.S. Global Change Research Program (NRC, 1999a). However, relatively little research has been done so far on the influence these climatic changes have on other environmental systems on the decade-to-century time scale. Physical geographers in general and USGS geographers in particular can contribute to the understanding of the impacts of climate change on other Critical Zone environmental systems. As with the study of climate the study of impacts of climate change will depend on exploitation of historical data. The results of historical studies can complement investigations of current processes. The U.S. Global Change Research Program has already proposed a number of key questions that are ripe for USGS contributions (USGS, 2001a).

The USGS should assign high priority to investigations of the following major questions:

- How do ecosystems respond to multiple stresses, such as combined climate change and human impacts?
- Given that climate changes of the past 300 years have occurred at the same time as rapid human modifications to ecosystems and land surface, what have been the relative and specific effects of climate change versus the effects of human activities?
- How will climate change affect biological diversity and ecosystem function?
- How will the spatial distribution of terrestrial species and biomes affect their biological and spatial responses to climatic change, and

what are the implications for forestry, biodiversity, and endangered species?
- How will climate change affect aquatic habitat and aquatic biodiversity?
- How will climate change affect water supply and quality, including contaminant transport?
- How will climate change affect geomorphic systems, including rivers and their floodplains, glaciers, coastal erosion, and the activity of sand dunes and wind-borne dust?
- How will abrupt climate change or increased extreme weather events affect processes that lead to natural hazards, such as landslides, floods, and wildfires?

The exploration of these questions depends on the use of historical geographic data and a spatial integrative approach that should characterize the research activities of the Geography Discipline.

Scale and Pattern in Ecosystems

Examples of significant USGS projects with important geographic components include wildland fire analysis and the study of invasive non-indigenous species. These projects show the importance of geographic scale and pattern in ecosystem analysis.

Over the last five years wildland fires have taken a large toll on ecosystems and the built environment, and have caused considerable economic disruption. Such losses are likely to increase because people are constructing more and more buildings in and near wildland areas, including developments of second homes. Additional questions for further USGS geographic investigation include:

- How does human landscape modification (primarily through roads, structures, and vegetation modification) influence the scale of wildland fires?
- How can wildland fire hazard maps, currently at 1-km resolution, be revised to improve resolution and include more data so they are useful for fire management?
- What scale of analysis can most efficiently meet prediction needs?

The study of invasive non-indigenous species also has strong geographic components (USGS, 2001b). An estimated 6,500 non-native plants and animals have become established in the United States, and some have had major impacts on ecosystem composition and dynamics. Invasive species impose substantial economic costs on agriculture, fisheries, forestry, water quality, and recreation. Prominent examples of costly non-indigenous species are red fire ants in the southern United States, Africanized honeybees, tamarisk in many western river systems, kudzu in the southeast, and cheatgrass in the western rangelands. The Geography Discipline should collaborate with other disciplines in addressing the following research questions on non-indigenous species:

- What are the mechanisms, rates, and spatial patterns of spread of invasive non-indigenous species?
- How do human activities contribute to the spread of non-indigenous species?
- How does the spatial pattern of natural and human-dominated landscapes affect the spread of non-indigenous species?
- Which landscape characteristics advance the spread of non-indigenous species, and which characteristics inhibit them?

The committee observes that studies of wildland fire and invasive species have essential geographic components, and the Geography Discipline should collaborate with relevant program areas within the USGS and with other federal agencies.

River Systems

The Geography Discipline has the potential to make direct contributions to and collaborate with others in river system analysis. River systems are important as water sources for human use, transportation, and hotspots of biodiversity. Because so much human activity tends to be concentrated along rivers, conflicts between human use, water quality, and ecological health are common. Many organizations inside and outside government are addressing these issues, but our scientific knowledge base is inadequate to inform current management and restoration actions. Construction and development projects move ahead without a clear understanding of their full implications. Interdisciplinary research on rivers, with geographers' participation, includes investigations into their interactive hydrologic, sedimentologic, chemical, and ecological processes. Key questions for future USGS river research include the following:

- What are the spatial interactions among fluvial processes, riparian vegetation, surface and subsurface flow, and human impacts?
- How do these interactions vary in different regions and types of rivers?
- How is water quality affected by spatial variation in human and natural processes?
- How are the physical, biological, and chemical integrity of rivers influenced by land uses near channels and in watersheds?
- How do ecological patterns and geographic variability affect the opportunities and costs for aquatic and riparian restoration?

The Geography Discipline should play a pivotal role in research on river systems at the USGS as an integrator of hydrologic, biologic, geologic, and geographic data, information, and knowledge. Knowledge of the geographic variation of control and response variables in river systems is key to understanding the behavior of the systems and their responses to human influences.

Hazards Research: Nature, Technology, and Security

An emerging emphasis within the USGS is the application of its data on Earth environments and Earth processes to analyze hazards and related issues. This newly developing emphasis on nature-society relationships has been a classic theme in geography outside the USGS for a century (Chapter 2), suggesting a convergence of the research agenda of the USGS and that of the broader geographic discipline. Professional geographers continue to make important research contributions regarding natural, technological, and disease hazards. Likewise, geographic analyses and geographic data are essential to maintaining and strengthening national security.

Natural Hazards

The systematic study of natural hazards in the United States originated with the work of geographer Gilbert White and his students and colleagues (e.g., Burton et al., 1978). Geographers investigate natural and technological hazards by exploring the complicated relationships between society and the natural environment with natural science data that is often collected and archived by the USGS. The mission statement of the USGS clearly states the agency's interest in hazards research.

Two examples illustrate how the Geography Discipline should participate in the investigations of natural hazards: the specific case of coastal hazards and the more general issue of vulnerability. Rising sea levels and accelerated coastal erosion have placed increasing numbers of citizens at risk (Heinz Center, 2000). At the same time human activities such as estuarine dredging, damming of coastal rivers, and construction on sensitive beach areas have created additional complications. Human occupation of coastal areas has stressed ecological communities, including endangered species. The USGS has a long-standing interest in coastal areas from the standpoints of monitoring and research and has developed partnerships with other agencies and programs that have responsibility for coastal zone science and management, for example, NOAA's National Ocean Service, the U.S. Army Corps of Engineers, and the FEMA. The Coastal and Marine Geology Program of the Geology Discipline is a comprehensive effort to understand the geologic processes that underlie all environmental systems at the continent-ocean interface (NRC, 1999b). The Biology Discipline is becoming more involved in coastal research and the Water Discipline plays a significant role through its monitoring and research on rivers that feed fresh water and sediment to the coastal zone. The Geography Discipline should also be involved by serving as an integrator of data, information, and knowledge, and as an authority on the spatial aspects of the coastal system.

A more general example of research potential for the Geography Discipline is the emerging science of vulnerability that focuses on the susceptibility of coupled human-environmental systems to perturbations (Cutter, 2001). Understanding vulnerability, as well as resistance to hazards, is based on multidisciplinary, place-based approaches that should be facilitated by the organization and mission of the USGS. Examples of the vulnerability of human systems to unusual or extreme events include the sensitivity of urban areas to earthquakes and floods. An example of successful human adaptation to hazards is the large-scale agricultural systems of south Florida that are largely resistant to severe hurricane damage (NRC, 1999b).

The Geography Discipline should participate in the active development of vulnerability science, because geography has an array of powerful integrative tools that are well designed to deal with regional-scale science and management problems. Outside the agency NOAA, the NSF, and the Department of Energy (DOE) actively promote vulnerability research, and they are candidates for partnerships with the Geography Discipline.

Technological Hazards and Public Health

Technological hazards relate to the built infrastructure and disposal of waste products, such as toxic and radioactive materials. The study of technological and natural hazards includes assessment of risk and emergency management. Technological hazards represent undesirable outcomes of human actions in the Critical Zone. The Geography Discipline, with its focus on the Critical Zone, should make substantial contributions to research that explains the occurrence of technological hazards, defines their distribution, and creates successful responses.

The USGS should strengthen science to improve public health and safety. The USGS currently works with the U.S. Department of Agriculture and the Centers for Disease Control and Prevention to better understand environmental factors that influence the geographic distribution of the mosquito-borne West Nile virus in the United States. Research to determine environmental factors that affect the location of disease vectors, identify changing patterns of vector populations and their habitats, and relate these patterns to human activity improve the nation's ability to predict and control vector-borne diseases. The spread of disease and disease vectors frequently relates to human activity and human alteration of the environment, so geographic researchers should contribute their experience at the interface of nature and society to strengthen this important research element.

The Geography Discipline should contribute to epidemiological studies by providing the spatial framework and understanding of spatial processes that explain the spread of diseases. The combination of medical expertise from agencies such as the National Institutes of Health and the Centers for Disease Control and Prevention with the geographic expertise in the Geography Discipline could produce powerful predictive models for decision makers. The USGS has a unique opportunity to explore disease, as it relates to geologic and biologic barriers and pathways.

Homeland Security

Homeland security is a new challenge for the Geography Discipline (though geographers outside the USGS have contributed to civil defense planning). Increased awareness of threats to our homeland has placed new demands on our knowledge of geographic patterns and processes. Often the most basic geographic information, such as DEMs, is important for public and private agencies that deal with these threats. Planning responses to attacks on individual sites requires accurate knowledge of location and setting. For example, after the September 11, 2001, attacks on New York and Washington, the Geography Discipline supplied users with 115,000 maps and several thousand remotely sensed images. At a larger scale an understanding of the geographic characteristics of resource distribution systems (i.e., for water, natural gas, oil, electricity) can aid in responses to system failures. The USGS has a critical role to play in response to security threats, not only by supplying descriptive data and maps but also in the analysis of these products. *The National Map*, when it becomes a reality, should serve as the basic data and knowledge source for responses.

Often scientific generalizations that apply to earthquakes, volcanic eruptions, and hurricanes also apply to terrorist attacks. For example, it is not possible to prevent or predict earthquakes, so as a nation we emphasize preparedness and the ability to supply quick relief in the form of information, transportation, rescue, and emergency services, followed by a redistribution of resources. The same general approach is called for in the aftermath of terrorist attacks, yet much of our focus is on prevention. The Geography Discipline should explore the parallels among these various hazards and design research to improve the national response to potential attacks. The USGS should implement a homeland security support system founded on the general principles used by the agency for dealing with natural hazards. The Geography Discipline should assume responsibility for being the nation's primary provider of geospatial data for homeland security, with plans to support the expedited provision of data in times of need.

SECONDARY PRIORITIES

Urban Dynamics

Urban areas are home to three-fourths of the U.S. population and an even larger fraction of the nation's economic activity. The rapid growth of cities in areas that were once sparsely populated, the expansion of urban population and activities into rural areas, and the widening social and economic divisions within cities have sharpened the issue of urban sustainability (NSF, 2000). Urban sustainability focuses on the questions such as: Can the nation's land, water, and biological resources sustain the lives and livelihoods of an increasingly urbanized population? What are the long-term consequences of current urban development patterns? These questions highlight the need for research on the processes of urban growth in the United States, and on the impacts of urban growth on people and environments.

The Geography Discipline activities in urban matters include extensive databases and the Urban Dynamics project. The Urban Dynamics Project contributes to geographic research on urban growth and change. Aimed at analyzing land use change in and around urban areas and the environmental consequences of rapid urbanization, the Urban Dynamics project represents a shift of focus for the USGS. Throughout most of its history the USGS neglected urban areas. The Urban Dynamics project generates maps showing the expansion of urbanized areas at the expense of surrounding rural land uses.

The maps generated by the Geography Discipline lead to the next level of science, modeling processes and change revealed by the map data. Urban modeling requires understanding human and environmental processes that affect the observed changes at various geographic scales. Most urban dynamics models used by the Urban Dynamics project (e.g., Clarke and Gaydos, 1998) incorporate a limited set of processes and constraints, such as slope and access to transportation, while ignoring the powerful social and economic forces that drive urban land use change (Knox, 1994). These processes can be linked to the shifting patterns of employment, infrastructure, services, and environmental resources. Although social and economic forces fall outside the USGS's traditional domain, accurate modeling of land use change and prediction of its environmental effects requires knowledge of such human forces. It is unlikely that the Geography Discipline will be able to develop a strong social and economic science component: therefore, the USGS should develop partnerships with researchers at universities and at federal, state, and tribal agencies.

Science in the Urban Dynamics project is predicated on a three-step process: create the data through mapping, model the processes revealed using GIScience, and predict future changes in the geography of U.S. cities using the models. The project, however, has achieved only the first step in this progression. The Geography Discipline has successfully collected the necessary data, but now it should proceed with the research involved in modeling and prediction to serve its clients. Effective prediction of urban change will require substantial input from social scientists in the urban research community. The Geography Discipline should also explore development of other similar mapping and research projects for other non-urban domains.

Regional and Place-Based Research

The recent reorganization of the USGS facilitates the development of integrative research to address multidisciplinary, integrative regional and place-based issues. The reorganization aids investigations at geographic scales defined by areas that are smaller than a USGS administrative region but may span parts of two or more states. This scale of analysis is regional, as traditionally defined within the discipline of geography. Regions defined on the basis of environmental or socioeconomic systems do not respect state boundaries, so the regional perspective is entirely appropriate for a federal science agency such as the USGS. The boundaries of the Critical Zone are apolitical.

The regional perspective has a long tradition in geography. Geographic questions about regions focus on defining them and understanding the differences and interactions among them. They also concentrate on understanding how cultural, economic, and environmental processes interact within discrete areas of the earth's surface. The geographer's interest in regions and regional studies declined from the 1960s through the 1980s, but revived in the 1990s. Place-based studies are effective in using the empirical qualitative data about people, societies, and their institutions, and at allowing identification of causes of events in specific places that do not conform to broad generalizations (Johnston, 1997). Broad and abstract generalizations about human societies intended to apply to most or all places have been unsuccessful because regional diversity persists even as capitalism, mass media, and globalization serve as homogenizing agents. Human geographers therefore have advocated a focus on place or "locality" in studying the interactions among political, social, and economic activities and structures. Many environmental, economic, and political problems occur at the regional scale level, making the region a particularly effective spatial scale for study. A region may, however, contain multiple cultures and communities of interest; moreover, regions interact with and are influenced by larger-scale processes.

At the same time as human geographers' interest in regions revived, other disciplines began adopting the region as a useful scale of analysis. Biologists and biogeographers, for example, define hierarchical systems of ecoregions, which are used widely in natural resource management (e.g., Abell et al., 2000; Omernik, 1987). Hydrologists use river basins as natural units of study to analyze water quality and water quantity problems. Legal and political challenges in natural resource policy have resulted in a move toward large bioregional assessments intended to provide a scientific basis for policy. Since 1990 federal, state, and local agencies have undertaken major bioregional assessments for management. Examples of defined management regions include the interior Columbia River Basin, the Sierra Nevada region in California and Nevada, the Everglades region of South Florida, and the southern California desert region.

The USGS also undertakes regional studies, which the agency calls focus areas. The USGS Central Region, for example, has identified five focus areas: Mountain West, Desert Southwest/U.S.–Mexico Border, Gulf Coast and Lower Mississippi, Missouri/Middle Mississippi, and the Great Plains.

The USGS shift to a more regional structure has created new place-based, interdisciplinary research opportunities. In the USGS Central Region, for example, place-based research has focused on urban areas and has integrated all the agency's disciplines. Study areas include the urban corridor along the east face of the Rocky Mountain Front Range, St. Louis–Kansas City–Omaha, Albuquerque, and Dallas–Ft. Worth. Urban growth and land use changes in these regions present challenges in providing natural resources such as gravel and aggregate, in protecting air and water quality and wildlife habitat, and in coping with natural hazards.

Other USGS administrative regions are also shifting toward place-based approaches. A high-priority regional research challenge in the Western Region is the integrated study of multiple natural hazards in the Pacific Northwest. The Western Region has also established an integrated science plan to study a set of rapidly growing urban areas (Tucson, Sacramento, Seattle, and Anchorage). Even before USGS's structural change geographers in the Eastern Region were involved in large integrated ecosystem research projects focusing on the Chesapeake Bay and the Everglades. The USGS Place-Based Studies program, established in 1995, seeks to provide objective-integrated science for managers seeking to restore natural functions and values of resources and the environment. The Place-Based Studies program began with two ecosystems—the San Francisco Bay and Delta, and South Florida—and added several others between 1995 and 1999. In each place research issues are generally focused on human influence on species and biodiversity, but emphasize the effect of water-quality decline on species. The USGS's Biology Discipline offers other examples of integrative regional research in its national status and trends reports (LaRoe et al., 1999; Mac and Opler, 1999) and the Land Use History of North America project.

The USGS should undertake three efforts to improve its regional and place-based geographic research. First, the Geography Discipline should learn how such research is conducted by the EPA in its Water and Watersheds program and by the NSF in its Biocomplexity Initiative. Second, Geography Discipline researchers should increase their interactions with academic geographers in professional meetings. Third, USGS researchers should extend their isolated regional studies to include comparisons between and among regions. As an ultimate objective USGS researchers should strive to become the integrative regional experts for the nation.

Bridging Science, Policy, and Decision Making

All the research topics identified in this chapter, as well as many aspects of geographic data and GIScience defined in Chapters 3 and 4, represent USGS activities that could build links among science, policy, and decision making. The most important arenas in which the agency should employ its expertise are the expansion of public involvement in decision making and the development of decision-support systems.

Expanding Public Involvement in Policy Making

Policy making increasingly involves many participants, including private citizens. Many policy decisions are made at regional and local levels as a result of decentralization of government during the last decade. Decentralization or devolution of policy making requires that the local communities receive adequate scientific information, in accessible form, to be able to make informed decisions. Community-based environmental management, for example, requires community-level data (Wondolleck and Yaffee, 2000). Most of the locally useful geography data and information are created and archived at the federal level. It is the responsibility of the Geography Discipline to deliver the right data and information to its regional clients in a timely fashion.

The trends toward decentralization and spatial disaggregation are dramatically increasing the demand for environmental data. Both the USGS and the EPA have been leaders among federal agencies in providing environmental information, much of which is now available on the Web. One key feature that makes these data useful for local decision making is the ability to search for data geospatially, using a map interface. However, despite apparent successes, challenges remain. How successful are these interfaces for users at various levels of expertise? Are different interfaces needed for some users? What simple analysis tools can or should be added to these data delivery systems? These questions should drive Geography Discipline research that can improve the dissemination of data and information to public users.

Decision-Support Systems

The development of links between scientific research and policy making presents a unique set of challenges, because scientific questions and findings are usually laden with complexity and detail. Decision makers often need generalizations, conclusions, or direct statements. Typically, scientists believe that the complexity of natural systems, as revealed through research, is an important lesson. An accurate portrayal of research results thus should not oversimplify complexity or abstract from it extensively, but for the decision maker simpler is easier, if not better.

In a similar vein, scientists are often reluctant to predict outcome of specific actions or policy decisions because scientific results are frequently characterized by uncertainty. In some natural systems, such as the atmosphere, responses to external perturbations may exhibit a threshold response or a shift between very different states, so differences in predictions may involve the identification of discrete states rather than simple changes in degree along a continuum of responses. By contrast policy makers usually want simple, direct statements of the likely outcome of a specific policy. Decision-support systems offer a valuable tool to bridge the complexity gap between researchers and decision makers. Decision-support systems are computer software tools that portray the dynamics of social and natural systems (Jankowski et al., 1997). Effective decision-support systems incorporate and display the data necessary for prediction and modeling. An important characteristic of decision-support systems is the capacity to display model results and predictions in ways that policy makers can clearly understand and evaluate. Invariably decisions about the environment have geospatial implications, therefore, decision-support systems have a GIS component, although the decision-support system is more than a GIS. Decision-support systems provide alternatives for multiple users to examine; however, they do not make decisions—people make decisions.

The USGS has produced several successful examples of decision-support systems. The Geospatial Multi-Agency Coordination Group (GeoMAC) is a decision-support system used by wildfire managers to manage and allocate resources for suppressing wildfires. It is a mapping tool USGS scientists developed in collaboration with the DOI and other agencies, including the National Interagency Fire Center. Developed in 2000–2001, it was used successfully to manage firefighting support in the Rocky Mountains during the extreme fire season of the summer of 2000. It allowed fire managers to anticipate rates of spread and identify critical locations where firefighting forces could be strategically placed. A second example of a successful decision-support system developed by USGS is the Habitat Needs Assessment Query Tool for the upper Mississippi River.Users of this tool can assess the impact of the U.S. Army Corps of Engineers' activities on aquatic organisms and migratory waterfowl. The USGS should pursue a number of research questions related to development of decision-support systems.

- How can data access be made more efficient and rapid?
- How can these systems be configured to handle larger and more diverse datasets as they are brought into geospatial form? Qualitative data are important in describing natural systems, as well as human systems.
- How can models use qualitative data in combination with quantitative data, and how can analysis and display of qualitative data be improved?
- How do policy makers typically use data and systems, and how can user interfaces be improved to fit the ways people make decisions?

- How can the flexibility of decision support systems be increased, so that a broader range of questions and types of scenarios can be posed?
- How can decision support systems be made more accessible and easier to use, so that these participants can use them effectively?

Geographic data often provide the mechanism for effective integration in complex projects or problems. Therefore, the Geography Discipline should lead integrative research on natural science and decision-support systems for the Critical Zone.

SUMMARY

The Geography Discipline contributes to the improvement of the quality of life of the nation's citizens by providing the data, information, and knowledge to deal with environmental resources, urban change, hazards, regional and place-based issues, and linking science with policy and decision making. The Geography Discipline excels at data management, the descriptive first step in science. The Geography Discipline also provides value-added components to convert the data to useful information. The Geography Discipline should not attempt to undertake research in all areas of geography but should focus on those topics close to its mission and those investigations that have a strong component of natural science. The Geography Discipline should now progress to the next steps in science: cutting-edge modeling and prediction that provide knowledge required by decision makers.

6

Conclusions and Recommendations

Previous chapters defined the role of the Geography Discipline in undertaking specific activities to enhance the Survey's role in geographic research. In this final chapter the committee presents its conclusions and recommendations. These conclusions and recommendations provide perspective and guidance for the USGS as it defines its strategic directions for the Geography Discipline. The essence of the conclusions and recommendations can be distilled into threads that run throughout this report:

- The Geography Discipline should engage in scientific research.
- The geographic research throughout the USGS should provide integrative science for investigations of the Critical Zone (i.e., the Earth's surface and near-surface environment that sustains nearly all terrestrial life [NRC, 2001b, p. 36]; Sidebar 1-1).
- The Geography Discipline should develop partnerships within the agency and with the broad field of geography outside the agency.
- The Geography Discipline should develop a long-term core research agenda that includes several projects of the magnitude of *The National Map*.

The Survey's guiding vision and mission are (USGS, 2000):

Vision. The USGS is a world leader in the natural sciences through its scientific excellence and responsiveness to society's needs.

Mission. The USGS serves the nation by providing reliable scientific information to:

- describe and understand the earth;
- minimize loss of life and property from natural disasters;
- manage water, biological, energy, and mineral resources; and
- enhance and protect quality of life.

The connection between geography and the USGS is the Survey's mission statement. Geography's interests in hazards research, resource analysis, connections between social and natural science, and integrative spatial methods make the discipline vital to the future success of the Survey.

A summary of the committee's conclusions and recommendations to achieve these contributions is organized according to the original charge to the committee. Specifically, the committee was charged to consider the following areas of concern to the USGS's Geography Discipline:

- The role of the USGS in advancing the general state of knowledge of the discipline (geography, cartography, GIScience);
- The role of the USGS in improving the understanding of the dynamic connections between the land surface and human interactions with it;
- The role of the USGS in maintaining and enhancing the tools and methods for conducting and applying geographic research; and
- The role of the USGS in bridging the gap between geographic science, policy making, and management.

The components of the charge ask the committee to assess the role of the USGS in scientific activities. One general answer is that the Survey's role in science is directly determined by the pattern of its culture and behavior:

- If the Survey conducts cutting-edge research in geography, then its research advances the science;
- If the Survey does not do cutting-edge research, but develops and disseminates tools and information products that help other researchers do their work, then the Survey indirectly advances the science;
- If the USGS funds external research by others, it influences the direction of geographic research; or
- If the USGS uses only the tools, information, and knowledge generated by others to carry out its mission, the Survey is merely a consumer and has no role in furthering the science.

The committee concludes that the Survey should follow the first three of these behavior patterns, but that at present the Survey does not exhibit an appropriate balance among them. The committee offers the following specific

CONCLUSIONS AND RECOMMENDATIONS

conclusions and recommendations to help the USGS achieve a balance among its activities. The recommendations in this chapter are broadly defined; the text of the preceding chapters contains specific and detailed recommendations.

ADVANCING THE GENERAL STATE OF KNOWLEDGE IN THE DISCIPLINE

The USGS can contribute to geographic knowledge by taking a leadership position in a few important areas of the field addressing the Critical Zone (NRC, 2001b, p. 36; Sidebar 1-1). It is uniquely suited to provide leadership in GIScience and remote sensing, but it can also contribute in other primary areas, including spatial analysis and nature-society interactions.

Conclusion: Currently the USGS's influence is weak in advancing the state of knowledge in general geography (i.e., geographic research other than GIScience) because Survey personnel conducting such research are not sufficiently engaged with geographers outside the Survey (Chapter 2).

Recommendation: To advance the state of knowledge in geography in general the USGS should strengthen its connections to the scientific community outside the Survey. These connections will be improved when Survey personnel participate in national geographic organizations and present USGS geographic research at professional geography meetings and in professional journals.

Conclusion: The USGS's influence is weak in advancing the state of knowledge in general geography because geographers at the Survey are limited to cartographic, geographic information systems (GIS) and remote sensing specialties, largely at the technical level (Chapter 2).

Recommendation: The USGS should expand its capabilities in geography beyond the activities of cartographic technicians to include leading-edge geographic research in GIScience, spatial analysis, and nature-society interactions.

UNDERSTANDING THE DYNAMICS OF THE LAND SURFACE–HUMAN ACTIVITIES CONNECTION

Part of the USGS role in advancing knowledge in geography includes specific contributions related to the Critical Zone. The Survey mission statement recognizes special responsibilities for hazards resulting from the interaction of society with nature.

Conclusion: The USGS manages large amounts of data to assess processes at the nature-society interface and provides a supporting mechanism for responses to natural disasters. Even though the academic field of geography is a significant contributor to the understanding of environmental processes and natural hazards, the Survey does not contribute greatly to the understanding of the vital connection between nature and society through scientific research focused on hazards (Chapter 5).

Recommendation: The USGS should continue to exercise national leadership in applied hazards research (including natural, technical, and security hazards) to improve the nation's explanatory, predictive, and response capabilities. To meet national needs, however, it is incumbent on the Survey to undertake basic research on environmental processes, hazards, and vulnerability, and to include the expertise of geographers and social scientists from within the Survey and through cooperative agreements.

Conclusion: The USGS manages and provides a variety of basic data for the nation's responses to natural and technical hazards. These data and methods of analysis are also applicable to issues related to homeland security, a subject that has many data and research similarities to investigations of natural and technical hazards (Chapter 5).

Recommendation: The USGS should implement a homeland security support system founded on the general principles used by the Survey for dealing with natural hazards.

MAINTAINING AND ENHANCING GEOGRAPHIC TOOLS AND METHODS

Traditionally the USGS maintained and enhanced tools and methods in geography to fulfill the Survey's role as one of the nation's primary sources of spatial data. Today it is not enough to supply data. In the digital era the nation requires new knowledge about how to deal with those data through GIScience. Consequently, the Survey's Geography Discipline should undertake basic research related to spatial data and develop cutting edge technologies in support of innovative geographic products such as *The National Map*. The role of the USGS should not stop with supplying data; it should

CONCLUSIONS AND RECOMMENDATIONS   103

continue with advanced research to develop new tools and methods to convert those data to information and ultimately to knowledge.

Conclusion: The USGS manages a national treasure of historic data ranging from maps and remotely-sensed imagery to long-term data collected from biologic, hydrologic, and geologic systems. These historic data are not artifacts valuable only for their curiosity. Rather, they indicate long-term trends in natural systems and baseline measures to assess human influences. Historic data allow the interpretation of present data, but use of the historic information is restricted by several unsolved problems related to access, processing, and analysis (Chapter 3).

Recommendation: The USGS should develop projects focused on historic data to address basic geographic research questions related to the accuracy, availability, quality, and scale issues for historical spatial data.

Conclusion: The terrorist attacks of September 11, 2001, raise issues regarding data security, especially for data the USGS manages, including imagery, maps, and water supply data. The Survey's data management responsibilities are conflicting. On one hand, one of the Survey's purposes is to make these data widely available; on the other, the federal government has a responsibility to protect data that might be used against the nation. At the USGS four associate directors determine which data to make available within their own disciplines. Because only general guidelines are available, the four associate directors' restrictions could be inconsistent (Chapter 3).

Recommendation: A uniform security policy for spatial data should be developed, and the associate directors should serve as advisors to a single USGS decision maker. To make as much data available as possible the policy should clearly state how the mission of the Survey and the security of the nation should be balanced in making decisions for data management.

Conclusion: Although Congress has designated the USGS as the clearing-house for spatial data, other Department of Interior (DOI) bureaus and federal agencies create and use spatial data. The underlying problem is a lack of integration among these geospatial databases (those databases with locational identifiers attached to data entries), which does not serve the scientific and public good. Addressing this problem requires research, standards, and the application of integrating methods (Chapter 3).

Recommendation: The USGS is ideally suited to be the lead agency in providing and managing spatial data, and the federal government should make available resources commensurate with the level of the task. The USGS should play a leading and facilitating role in shaping national policy on geospatial data and developing an interoperable capability that will make it a primary access point for integrated geospatial data in the Department of the Interior and other federal agencies

Conclusion: *The National Map* is a bold vision for the future of the Geography Discipline, with the spatial database of the same name being its most prominent product. Without question the digital era has made the paper topographic map series obsolete for many applications, but *The National Map* will not become a reality with the present level of knowledge about the tools and methods needed to create the product (Chapter 4).

Recommendation: Given the importance of *The National Map* to the information economy of the future, and the need for further supportive research to accomplish *The National Map*, the Geography Discipline's programs—Cooperative Topographic Mapping, Land Remote Sensing, and Geographic Analysis and Monitoring—should receive a level of funding commensurate with the task.

Conclusion: Construction and maintenance of *The National Map* will require a variety of databases, but some are of exceptional priority if *The National Map* is to succeed. These high-priority datasets will require emphasis in funding and support (Chapter 4).

Recommendation: Because of their importance to *The National Map*, the following datasets should be assigned the highest priority in distribution of resources and in establishing and improving interagency exchanges:

- orthorectified imagery;
- digital elevation data;
- land cover data;
- biogeographic data;
- hydrographic data;
- transportation feature data; and
- geographic place names.

Conclusion: The USGS is effective at creating and managing spatial data, but its role in GIScience is limited and does not include cutting-edge research in geographic information systems or the analysis of the data that the Survey provides to others. The Survey has a weak core research program in geographic science related to the discipline's tools and methods (Chapter 4).

Recommendation: To achieve the vision and mission of the USGS, the Survey should improve its contributions to geographic knowledge, tools, and techniques by developing the internal capability to address the high-priority subjects of:

- resolution and scale;
- delivering vector data to users;
- standards for spatial data; and
- spatial statistics and analysis.

BRIDGING THE GAP BETWEEN SCIENCE, POLICY MAKING, AND MANAGEMENT

The mission of the USGS includes investigations in the Critical Zone to enhance and protect the quality of life, and contribute to wise development. The Survey already provides valuable service to its partners and clients by supplying spatial data and information, but the appropriate role of the Survey in general and the Geography Discipline in particular includes fundamental research. The integrative power of geographic analysis and the communications power of geographic data can be substantially enhanced through research conducted by the Geography Discipline.

Conclusion: The USGS is regionalizing its activities. This development positions the Survey to contribute to regional research and policy activities. To capitalize on this transformation the USGS should conduct substantive research that is explicitly regional, place-based, and integrated, rather than more traditional topically defined biologic, hydrologic, and geologic investigations (Chapter 5).

Recommendation: The USGS should strengthen its regional and place-based research (as opposed to topically divided investigations in geology, hydrology, and biology), including extensive involvement with regional research outside the Survey. The USGS should develop the ability to provide integrative regional experts for the nation.

Conclusion: The USGS cannot address all problems associated with bridging science, policy, and decision making, but its Geography Discipline can lead research activities in a few priority areas likely to draw upon existing expertise in the field of geography and improve the bridging function. GIS and remotely-sensed products promote citizen involvement at public meetings by providing a mode of communication between specialist and layperson based on data, while place-based frameworks and decision-support systems allow for experimentation to assist decision makers. Currently, the Survey lacks substantial research capability in these priority areas (Chapter 5).

Recommendation: The USGS should assign high priority and substantial resources to fundamental research directed toward:

- improving citizen involvement in decision making for issues related to natural sciences by creating citizen-friendly geographic interfaces with all the Survey's primary spatial datasets;
- expanding the utility and application of place-based science by conducting integrative place-specific research in addition to topical research in individual disciplines; and
- enhancing the effectiveness of decision-support systems with increased geographic input and more effective map-like products as output.

SUMMARY

The USGS is reforming and incorporating missions that emphasize its role as one of the nation's most important natural science research agencies. The Geography Discipline produces valuable spatial data for users ranging from private citizens and corporations to governmental agencies at all levels. The Geography Discipline should now expand its activities to assume its proper role among the other disciplines at the USGS by engaging in fundamental geographic research, investigating the processes and forms that explain the dynamics of location, space, and place. The investment in such research will change the Geography Discipline, but it will pay enormous dividends for the nation by improving the science done in other disciplines, integrating new knowledge and data generated by the USGS and others, reducing losses from hazards, improving management of natural resources, enhancing the quality of life, and aiding in wise development. A strong Geography Discipline with a productive research component will ensure recognition of the USGS as scientifically credible, objective, and relevant to society's needs.

References

Abell, R. A., D. M. Olson, E. Dinerstein, P. T. Hurley, J. T. Diggs, W. Eichbaum, S. Walters, W. Wettengel, T. Allnutt, C. J. Loucks, and P. Hedao. 2000. *Freshwater Ecoregions of North America: A Conservation Assessment.* Washington, D.C.: World Wildlife Fund-United States. 368 pp.

Anderson, J. R., E. E. Hardy, J. T. Roach, and R. E. Witmer. 1976. *A Land Use and Land Cover Classification System for Use with Remote Sensor Data.* U.S. Geological Survey Professional Paper 964. Washington, D.C.: U.S. Government Printing Office. 28 pp.

Burton, I., R. W. Kates, and G. F. White. 1978. *The Environment as Hazard.* New York, New York: Oxford University Press. 240 pp.

Buttenfield, B. P. 1997. The Future of the Spatial Data Infrastructure: Delivering Geospatial Data. *GeoInfo Systems*, June, pp. 18-21.

Buttenfield, B. P. 1999. Sharing Vector Geospatial Data on the Internet. *Proceedings, 18th Conference of the International Cartographic Association*, August, Ottawa, Canada, Section 5: 35-44. Accessed on December 12, 2001, at: <http://greenwich.colorado.edu/ babs/ottawa/ottawa.htm.>

Buttenfield, B. P. 2001. Mapping Ecological Uncertainty. Pp. 115-132 in C. Hunsaker, M. F. Goodchild, M. Friedl, and T. Case, eds. *Uncertainty in Spatial Data for Ecological Analyses.* New York, New York: Springer-Verlag.

Clarke, K. and L. Gaydos. 1998. Loose-coupling a Cellular Automaton Model and GIS: Long-term Urban Growth Prediction for San Francisco and Washington/Baltimore. *International Journal of Geographical Information Science*, 12(7): 699-714.

Colwell, R. N., ed. 1983. *Manual of Remote Sensing*, 2nd Edition. Falls Church, Virginia: American Society of Photogrammetry. 2440 pp.

Curran, P. J. 1987. Remote Sensing Methodologies and Geography. *International Journal of Remote Sensing*, 8:1255–1275.

Cutter, S., ed. 2001. *American Hazardscapes: Regionalization of Hazards and Disasters.* Washington, D.C.: Joseph Henry Press. 211 pp.

DOI [U.S. Department of the Interior]. 2001. Protection of Sensitive Information. Letter from P. Lynn Scarlett, assistant secretary for policy, management, and budget, to solicitor, inspector general, assistant secretaries, and heads of bureaus and offices, November 8.

Dahlberg, R. E. and J. R. Jensen. 1986. Education for Cartography and Remote Sensing in the Service of an Information Society. *The American Cartographer,* 13(1):51-71.

Draeger, W., T. M. Holm, D. T. Lauer, and R. J. Thompson. 1997. The Availability of Landsat Data: Past, Present and Future. *Photogrammetric Engineering & Remote Sensing,* 63(7):869-875.

Eckel, E. B. 1982. *The GSA: Life History of a Learned Society.* GSA Memoir 155. Boulder, Colorado: Geological Society of America. 167 pp.

Egan, D. and E. A. Howell. 2001. *The Historical Ecology Handbook: A Restorationist's Guide to Reference Ecosystems.* Washington, D.C.: Island Press. 457 (+xix) pp.

Ester, M., H. P. Kriegel, and J. Sander. 1997. Spatial Data Mining: A Database Approach. *Lecture Notes in Computer Sciences,* 1262: 47-66.

Fosnight, E. 1992. Data Integration through Region-based Nominal Filtering. *International Journal of Geographical Information Systems,* 6(6): 469-478.

Geography Education Standards Project. 1994. *Geography or Life: National Geography Standards.* Washington, D.C.: National Geography Society. 272 pp.

Heinz Center. 2000. *The Hidden Costs of Coastal Hazards.* Washington, D.C.: Island Press. 252 pp.

James, P. E. and G. J. Martin. 1978. *The Association of America Geographers: The First Seventy-five Years, 1904-1979.* Washington, D.C.: Association of American Geographers. 279 pp.

Jankowski, P., T. L. Nyerges, A. Smith, T. J. Moore, and E. Horvath. 1997. Spatial Group Choice: A SDSS Tool for Collaborative Spatial Decision-Making. *International Journal of Geographic Information Science,* 11(6): 577-602.

Jensen, J. R. 2000. *Remote Sensing of the Environment: An Earth Resource Perspective.* Saddle River, New Jersey: Prentice-Hall, Inc. 544 pp.

Jensen, J. R., D. Cowen, X. Huang, D. Graves and K. He. 1996. Remote Sensing Image Browse and Archival Systems. *Geocarto International - A Multidisciplinary Journal of Remote Sensing & GIS,* 11(2):33-42.

Johnston, R. J. 1997. *Geography and Geographers: Anglo-American Human Geography since 1945,* 5th edition. London, U.K.: Arnold Publishing. 304 pp.

Kelmelis, J.A. 2001. Geographic Mapping and Analysis at the United States Geological Survey. Pp. 4: 2208-2217 in *Mapping the 21st Century: Proceedings of the 20th ICA/AIC International Cartographic Conference,* Beijing, China, August 6-10, 2001. Bejing: Bejing Press of Surveying and Mapping.

Knox, P. 1994. *Urbanization: An Introduction to Urban Geography.* Englewood Cliffs, New Jersey: Prentice Hall, Inc. 436 pp.

LaRoe, E. T., G. S. Farris, C. E. Puckett, P. D. Doran, and M. J. Mac, eds. 1999. *Our Living Resources: A Report to the Nation on the Distribution, Abundance, and Health of U.S. Plants, Animals, and Ecosystems.* Washington, D.C.: National Biological Service. 530 pp.

REFERENCES

Longley, P. A., M. F. Goodchild, D. J. Maguire, and D. Rhind. 1999. *Geographic Information Systems and Science: Principles, Techniques, Management and Applications*, 2nd edition. Chichester, U.K.: Wiley. 1101 (+xciii) pp.

Longley, P. A., M. F. Goodchild, D. J. Maguire, and D. Rhind. 2001. *Geographic Information Systems and Science*. Chichester, U.K.: Wiley. 454 (+18) pp.

Mac, M. J. and P. A. Opler, eds. 1999. *Status and Trends of the Nation's Biological Resources*. Washington, D.C.: U.S. Geological Survey, Biological Resources Discipline. Accessed April 15, 2002, at: <http://biology.usgs.gov/s+t/SNT/index.htm>.

Miller, H. and J. Han, eds. 2001. *Geographic Data Mining and Knowledge Discovery*. London, U.K.: Taylor & Francis. 372 pp.

NAPA [National Academy of Public Administration]. 1998. *Geographic Information for the Twenty-First Century: Building a Strategy for the Nation*. Washington, D.C.: National Academy of Public Administration. 358 pp.

NIST [National Institute of Standards and Technology]. 1994. Federal Information Processing Standard Publication 173 (Spatial Data Transfer Standard Part 1, Version 1.1). Washington, D.C.: Department of Commerce. Accessed on April 16, 2002, at: <http://www.itl.nist.gov/fipspubs/fip173-1.pdf>.

NRC [National Research Council]. 1993. *Distributed Geolibraries: Spatial Information Resources*. Washington, D.C.: National Academy Press. 119 pp.

NRC [National Research Council]. 1994. *Promoting the National Spatial Data Infrastructure through Partnerships*. Washington, D.C.: National Academy Press. 128 pp.

NRC [National Research Council]. 1995. *A Data Foundation for the National Spatial Data Infrastructure*. Washington, D.C.: National Academy Press. 45 pp.

NRC [National Research Council]. 1997. *Rediscovering Geography: New Relevance for Science and Society*. Washington, D.C.: National Academy Press. 234 pp.

NRC [National Research Council]. 1999a. *Global Environmental Change, Research Pathways for the Next Decade*. Washington, D.C.: National Academy Press. 82 pp.

NRC [National Research Council]. 1999b. *Science for Decisionmaking: Coastal and Marine Geology at the U. S. Geological Survey*. Washington, D.C.: National Academy Press. 124 pp.

NRC [National Research Council]. 2001a. *Basic Research Opportunities in Earth Science*. Washington, D.C.: National Academy Press. 153 pp.

NRC [National Research Council]. 2001b. *Future Roles and Opportunities for the U.S. Geological Survey*. Washington, D.C.: National Academy Press. 179 pp.

NSF [National Science Foundation]. 2000. *Towards a Comprehensive Geographical Perspective on Urban Sustainability*. New Brunswick, New Jersey: Center Urban Policy Research Press. 29 pp.

Oliver, C. D., L. L. Irwin, and W. H. Knapp. 1994. Eastside Forest Management Practices: Historical Overviews, Extent of their Application, and their Effects on Sustainability of Ecosystems. In P. F. Hessburg, ed., *Eastside Forest Ecosystem Health Assessment*, Volume III. Wenatchee, Washington: USDA Forest Service, National Forest System. 74 pp.

Omernik, J. M. 1987. Ecoregions of the Conterminous United States. *Annals of the Association of American Geographers,* 77:118-125.

Pyne, S. J. 1980. *Grove Karl Gilbert: A Great Engine of Research.* Austin, Texas: University of Texas Press. 306 pp.

Quattrocchi, D. and M. F. Goodchild, eds. 1997. *Scale in Remote Sensing and GIS.* Boca Raton, Florida: CRC Press. 406 pp.

Rabbitt, M. C. 1980. *Minerals, Lands, and Geology for the Common Defense and General Welfare,* Volume 2, 1879-1904. Washington, D.C.: U.S. Geological Survey and U.S. Government Printing Office. 407 pp.

Salge, F. 1995. Semantic Accuracy. Pp. 139-152 in S. C. Guptill and J. L. Morrison, eds. *Elements of Spatial Data Quality.* Oxford, U.K.: Elsevier.

Seaber, P. R., F. P. Kapinos, and G. L. Knapp. 1987. Hydrologic Units Maps. USGS Water-Supply Paper 2294. Denver, Colorado: USGS Information Services. 63 pp.

Stegner, W. 1954. *Beyond the Hundredth Meridian: John Wesley Powell and the Second Opening of the West.* Boston, Massachusetts: Houghton Mifflin. 438 pp.

Thompson, M. M. 1981. *Maps for America: Cartographic Products of the U.S. Geological Survey and Others*, 2nd edition. Washington, D.C.: U.S. Government Printing Office. 265 pp.

Thrower, N. J. and J. R. Jensen. 1976. The Orthophoto and Orthophotomap: Characteristics, Development, and Aspects of Cartographic Communication. *The American Cartographer,* 3(1):39-56.

UCGIS [University Consortium of Geographic Information Science]. 1998. *Spatial Analysis in a GIS Environment.* Washington, D.C.: University Consortium on Geographic Information Science. White Paper on Emerging Research Themes. Accessed December 4, 2001, at: <http://www.ucgis.org/research_white/anal.html>.

USGS [U.S. Geological Survey]. 1996. *Strategic Plan for the U.S. Geological Survey–1996-2005.* Reston, Virginia: U.S. Geological Survey. 43 pp.

USGS [U.S. Geological Survey]. 1999. *One Bureau, One Mission, One Message. Strategic Change of the U.S. Geological Survey.* Reston, Virginia: U.S. Geological Survey.

USGS [U.S. Geological Survey] 2000. *U.S. Geological Survey Strategic Plan 2000-2005.* Reston, Virginia: U.S. Geological Survey. 25 pp. Accessed July 23, 2002, at: <http://www.usgs.gov/stratplan/stratplan_rev.pdf>.

USGS [U.S. Geological Survey]. 2001a. *National Mapping Program International Activities, Current and Future. National Mapping Program International Activities for the U.S. Department of State.* Reston, Virginia: U.S. Geological Survey. Accessed on July 29, 2002, at: <http://mapping.usgs.gov/html/international/international.html>.

USGS [U.S. Geological Survey]. 2001b. *The National Map Web site.* Accessed August 12, 2002, at: <http://nationalmap.usgs.gov/>.

USGS [U.S. Geological Survey]. 2001c. *Earthshots: Satellite Images of Environmental Change.* Accessed on April 12, 2002, at: <http://edcwww.cr.usgs.gov/earthshots/slow/tableofcontents>.

USGS [U.S. Geological Survey]. 2001d. *Aerial Photographs and Satellite Images.* Illustrated pamphlet. Reston, Virginia: U.S. Geological Survey. 20 pp.

REFERENCES

USGS and EPA [U.S. Geological Survey and U.S. Environmental Protection Agency]. 2000. *The National Hydrography Dataset: Concepts and Content.* Accessed Oct. 20, 2001, at: <http://nhd.usgs.gov/techref.html>.

Vogelmann, J. E., S. M. Howard, L. Yang, C. R. Larson, B. K. Wylie, and J. N. Van Driel. 2001. Completion of the 1990s National Land Cover Data Set for the Conterminous United States. *Photogrammetric Engineering and Remote Sensing*, 67:650-662.

Witmer, R. 2000. *Geography and Geographers at the USGS: A Historical Perspective.* Presentation to Committee on Geography, National Research Council, Washington, D.C., May 24.

Wondolleck, J. M., and S. L. Yaffee. 2000. *Making Collaboration Work: Lessons from Innovation in Natural Resource Management.* Washington, D.C.: Island Press. 277 pp.

Wortman, K. and B. P. Buttenfield. 1994. Preface. *Cartography and GIS,* 21(3):131. (Special Issue: Current Developments and Use of the Spatial Data Transfer Standards.)

Yuan, M., B. P. Buttenfield, M. Gahegan, and H. Miller. 2001. Geospatial Data Mining and Knowledge Discovery. *UCGIS White Paper.* Accessed on April 15, 2002, at: <http://www.ucgis.org/emerging/gkd.pdf>.

Appendixes

Appendix A

Biographical Sketches of Committee Members and Staff

William L. Graf (Chair) is Educational Foundation University Professor and Professor of Geography at the University of South Carolina. His research interests focus on the physical forms and processes associated with rivers, and on policy for public land and water. He is a past president of the Association of American Geographers. He has received more than 50 grants and contracts, and has authored or edited 7 books and more than 100 papers and book chapters in the fields of geography, geology, hydrology, public policy, and environmental history. His present research emphasis is on the physical environmental effects of large dams, and he is an advisor to federal agencies on decision making regarding dam decommissioning. He is a member of the NRC Board on Earth Sciences and Resources and has been a member of the NRC Water Science and Technology Board. He has participated in numerous NRC studies, and recently chaired the NRC Committee on Innovative Watershed Management.

Barbara P. Buttenfield is Professor of Geography at the University of Colorado in Boulder and Director of the Meridian Lab, a research facility focusing on visualization and modeling of geographic information and technology. She teaches courses in Geographic Information Science, Computer Cartography, and Geographic Information Design. She spent a one-year research sabbatical in residence at the United States Geological Survey (USGS) National Mapping Division in Reston, Virginia (1993-1994). Her research interests focus on data delivery on the Internet, visualization tools for environmental modeling, map generalization, and interface usability testing. Dr. Buttenfield is a Past President of the American Cartographic Association, and a fellow of the American Congress on Surveying and Mapping (ACSM). She was a member of the National Research Council's Mapping Science Committee 1992 to 1998.

Carol Harden is Professor of Geography at the University of Tennessee. She has been on the geography department faculty at the University of Tennessee since 1987, and served as Department Head from 1995 to 2000. Her research interests include process geomorphology, emphasizing hillslope processes and watershed dynamics in mountain regions; human impact on the environment, focusing on water; land use, and soil erosion relationships; and the geomorphology of deglaciated and deglaciating landscapes. Her professional memberships include the Association of American Geographers, the American Geophysical Union, and the American Water Resources Association. Dr. Harden served as chair of the Geo-morphology Specialty Group of the Association of American Geographers in 1997 to 1998 and as the U.S. representative to the International Association of Geomorphologists in 2002.

John R. Jensen is a Carolina Distinguished Professor in the Department of Geography at the University of South Carolina. He majored in physical geography and analytical cartography and remote sensing at the following institutions: B.A., California State University, Fullerton, 1971; Master's degree, Brigham Young University, 1972; and Ph.D., from the University of California, Los Angeles, 1976. He has mentored 50 master's students and 22 Ph.D.'s in remote sensing. Dr. Jensen's research focuses on: (a) remote sensing of vegetation biophysical resources (biomass, leaf-area-index); (b) developing improved digital image processing algorithms to extract and model change; (c) improvement of environmental sensitivity index (ESI) mapping to protect coastal resources; and (d) modeling water quality parameters (chlorophyll, dissolved inorganic matter) in estuaries and reservoirs using remote sensor data. He is past president of the American Society for Photogrammetry & Remote Sensing and author of *Introductory Digital Image Processing* (2nd Edition, 1996) and *Remote Sensing of the Environment* (2000) published by Prentice-Hall, Inc.

George P. Malanson is Professor of Geography at the University of Iowa, where he has been a Faculty Scholar and Intergraph Professor of Landscape Ecology. His research interests include biogeography and landscape ecology. His previous positions include terms at Oklahoma State University; CNRS Centre Emberger, Montpellier; and Southwest Texas State University. He is a past chair of the Biogeography Specialty Group of the Association of American Geographers and was a task leader in the International Geosphere-Biosphere Programme's Global Change in Terrestrial Ecosystems project. He is currently participating in a field research project on the invasibility of alpine tundra by woody vegetation funded through the USGS Glacier Field Station.

Patricia F. McDowell is Professor of Geography and Professor of Environmental Studies at the University of Oregon. She teaches courses in fluvial geomorphology, watershed science and policy, arid lands geomorphology, and soils geography. Her research focuses on response of river systems to human impacts and environmental change, and it has been supported by funding from the National Science Foundation, Environmental Protection Agency, U.S. Forest Service and Bonneville Power Administration. At the University of Oregon, she served as Associate Vice-President for Research in 1990 to 1993 and as Chair of the Department of Geography in 1993 to 1996. She was Chair of the Geomorphology Specialty Group of the Association of American Geographers in 1990-1991, and a Natural Science Research Council Guest Scientist at the Department of Physical Geography, Uppsala University, Sweden in 1991.

Sara McLafferty is Professor of Geography at the University of Illinois at Urbana-Champaign. Her research expertise centers on the application of geospatial techniques to the study of urban geography, specifically the use of spatial analysis methods and geographic information systems to analyze health, social, and environmental problems in cities. She served as a national councilor for the Association of American Geographers and was a member of the National Research Council's Mapping Science Committee.

Risa Palm is Professor of Geography and Dean of the College of Arts and Sciences at the University of North Carolina at Chapel Hill. Her research interests include societal responses to earthquake hazards and intra-urban mobility. She has done bi-national research on earthquake hazard response and preparation in California and southern Japan. Her previous positions include terms as Dean of the College of Arts and Sciences at the University of Oregon and Dean of the Graduate School and Associate Vice-Chancellor for Research at the University of Colorado at Boulder. She is a past president of the Association of American Geographers.

Norbert P. Psuty is Professor II in the Departments of Geography, Geological Sciences, and Marine and Coastal Sciences at Rutgers, The State University of New Jersey. He is also Associate Director of the Institute of Marine and Coastal Sciences. His coastal geomorphologist research encompasses beach and dune processes and morphology, sediment budget studies, barrier island dynamics, estuarine sedimentation, and sea-level rise. His research has been conducted primarily in various portions of coastal

New Jersey and it has both a basic science component as well as an applied side. He is a consultant to the National Park Service on shoreline dynamics and change in the coastal parks. He has been the Chair of the Coastal Commission of the International Geographical Union, President of The Coastal Society, and President of the New Jersey Academy of Sciences. He is recipient of the Honors Award from the Association of American Geographers.

Henry J. Vaux, Jr. is Associate Vice-President for Agriculture and Natural Resources (ANR) of the University of California, System-wide. He is also Professor of Resource Economics at the University of California, Riverside. His principal research interests are in the economics of water use and water quality. He also serves as Co-Chair of the Rosenberg International Forum on Water Policy, an ongoing global dialogue to reduce water-related conflict and improve water policy. He previously served as Director of the University of California Water Resources Center. Prior to joining the University of California he worked at the Office of Management and Budget and served on the staff of the National Water Commission. He received a Ph.D in economics from the University of Michigan. Dr. Vaux is immediate past Chair of the Water Science and Technology Board of the National Research Council and is a National Associate of the National Academies.

NRC Staff

Anthony R. de Souza is currently director of the Board on Earth Sciences and Resources at the National Research Council in Washington, D.C. Previously he was executive director of the National Geography Standards Project, secretary general of the 27th International Geographical Union Congress, editor of *National Geographic Research & Exploration*, and editor of the *Journal of Geography*. He has held positions as a professor and as a visiting teacher and scholar at the George Washington University, University of Wisconsin-Eau Claire, University of Minnesota, University of California-Berkeley, and University of Dar es Salaam in Tanzania. He has served as a member of NRC committees. He holds B.A. (honors) and Ph.D. degrees from the University of Reading in England, and has received numerous honors and awards, including the Medalla al Benito Juarez in 1992 and the Gilbert Grosvenor honors award from the Association of American Geographers in 1996. His research interests include the processes and mechanisms of economic development and human-environment relationships. He has published several books and more than 100 articles, reports, and reviews.

Lisa M. Vandemark has a Ph.D. in Geography from Rutgers University and a M.S. in Human Ecology from the University of Brussels, Belgium. Her B.S. (nursing, specialty psychiatry) is also from Rutgers University. Currently Lisa is a Program Officer at the National Research Council. Prior to this appointment she was a research associate at the Institute of Marine and Coastal Sciences, Rutgers University, and an intern at the National Science Resources Center at the Smithsonian Institution. Her research interests include environmental perception and decision making, international development, natural resource management, and the role of interdisciplinary studies in environmental protection.

Monica R. Lipscomb is a research assistant for the NRC Board on Earth Sciences and Resources. She is completing a Master's in Urban and Regional Planning at Virginia Polytechnic Institute, with a concentration in Environmental Planning and graduate certificate in International Development. Her areas of interest include community-based environmental protection, watershed management and environmental conflict resolution. Previously she served as a Peace Corps volunteer in Côte d'Ivoire and has worked as a biologist at the National Cancer Institute. She holds a B.S. in environmental and forest biology from the State University of New York–Syracuse.

Appendix B

Oral and Written Contributors

G. Bryan Bailey, USGS EROS Data Center, Sioux Falls, South Dakota

Jesslyn Brown, USGS EROS Data Center, Sioux Falls, South Dakota

William Carswell, U.S. Geological Survey, Denver, Colorado

Thomas Casadevall, U.S. Geological Survey, Denver, Colorado

Mark DeMulder, U.S. Geological Survey, Reston, Virginia

Frank D'Erchia, U.S. Geological Survey, Denver, Colorado

Thomas DiNardo, U.S. Geological Survey, Denver, Colorado

Chris Doescher, USGS-EROS Data Center, Sioux Falls, South Dakota

Mark Drummond, U.S. Geological Survey, Fort Collins, Colorado

Wynn Duong, U.S. Geological Survey, Reston, Virginia

John Dwyer, USGS EROS Data Center, Sioux Falls, South Dakota

Martin Eckes, U.S. Geological Survey, Reston, Virginia

Max Ethridge, U.S. Geological Survey, Rolla, Missouri

Joan Fitzpatrick, U.S. Geological Survey, Denver, Colorado

Thomas Fouch, U.S. Geological Survey, Denver, Colorado

Alisa Gallant, USGS EROS Data Center, Sioux Falls, South Dakota

Dean Gesch, USGS EROS Data Center, Sioux Falls, South Dakota

David Greenlee, USGS EROS Data Center, Sioux Falls, South Dakota

Susan Greenlee, USGS EROS Data Center, Sioux Falls, South Dakota

Charles Groat, U.S. Geological Survey, Reston, Virginia

David Hester, U.S. Geological Survey, Denver, Colorado

Robert Hirsch, U.S. Geological Survey, Reston, Virginia

Thomas Holm, USGS EROS Data Center, Sioux Falls, South Dakota

Collin Homer, USGS EROS Data Center, Sioux Falls, South Dakota

John Kelmelis, U.S. Geological Survey, Reston, Virginia

Patrick Leahy, U.S. Geological Survey, Reston, Virginia

John LaBrecque, NASA Headquarters, Washington, D.C.

Zhong Lu, USGS EROS Data Center, Sioux Falls, South Dakota

James Lusby, National Imagery and Mapping Agency, Bethesda, Maryland

Bonnie McGregor, U.S. Geological Survey, Reston, Virginia

William Miller, U.S. Geological Survey, Reston, Virginia

Carol Mladinich, U.S. Geological Survey, Denver, Colorado

Marilyn Myers, U.S. Geological Survey, Denver, Colorado

Maury Nyquist, U.S. Geological Survey, Denver, Colorado

Sara Jean Paulson, USGS EROS Data Center, Sioux Falls, South Dakota

Bruce K. Quirk, USGS EROS Data Center, Sioux Falls, South Dakota

Wayne Rohde, USGS EROS Data Center, Sioux Falls, South Dakota

Barbara Ryan, U.S. Geological Survey, Reston, Virginia

Karen Siderelis, U.S. Geological Survey, Reston, Virginia

Daniel Steinwand, USGS EROS Data Center, Sioux Falls, South Dakota

June Thormodsgard, USGS EROS Data Center, Sioux Falls, South Dakota

Kristine Verdin, USGS EROS Data Center, Sioux Falls, South Dakota

Raymond Watts, U.S. Geological Survey, Fort Collins, Colorado

Zhi-Liang Zhu, USGS EROS Data Center, Sioux Falls, South Dakota

Appendix C

Acronyms

AVHRR	Advanced Very High Resolution Radiometer
BGN	U.S. Board on Geographic Names
BTS	Bureau of Transportation Statistics
CDC	Centers for Disease Control and Prevention
CTM	Cooperative Mapping
DAAC	Distributed Active Archive Center
DEM	Digital Elevation Model
DLG	Digital Line Graphs
DOC	Department of Commerce
DOD	Department of Defense
DOE	Department of Energy
DOI	Department of Interior
DRG	Digital Raster Graphics
EPA	U.S. Environmental Protection Agency
EROS	Earth Resources Observation Systems
ETM+	Enhanced Thematic Mapper Plus
FGDC	Federal Geographic Data Committee
FIPS	Federal Information Processing Standard
FY	Fiscal Year
GAM	Geographic Analysis and Monitoring
GAP	Gap Analysis Program
GeoMAC	Geospatial Multi-Agency Coordination Group
GIS	Geographic Information Systems
GIScience	Geographic Information Science
GNIS	Geographic Names Information System
GPS	Global Positioning System

GS	General Schedule
HRV	Historical Range Of Variation
InSAR	Interferometric Synthetic Aperture Radar
IGBP	International Geosphere-Biosphere Programme
LIDAR	Light Detection and Ranging
LP DAAC	Land Processes Distributed Active Archive Center
LRS	Land Remote Sensing
MSS	Multispectral Scanner
NALC	North American Landcover
NASA	National Aeronautics and Space Administration
NDEP	National Digital Elevation Program
NED	National Elevation Dataset
NHD	National Hydrography Dataset
NIH	National Institutes of Health
NIMA	National Imagery and Mapping Agency
NLCD	National Land Cover Dataset
NMFS	National Marine Fisheries Service
NOAA	National Oceanic and Atmospheric Administration
NRC	National Research Council
NSDI	National Spatial Data Infrastructure
NSF	National Science Foundation
NWIS	National Water Information Service
SCT	Strategic Change Team
SDTS	Spatial Data Transfer Standard
TM	Thematic Mapper
USDA	U.S. Department of Agriculture
USGS	U.S. Geological Survey
UTM	Universal Transverse Mercator Grid

Appendix D

United States Department of the Interior

OFFICE OF THE SECRETARY
Washington, D.C. 20240

Memorandum   NOV - 8 2001

To: Solicitor
    Inspector General
    Assistant Secretaries
    Heads of Bureaus and Offices

From: P. Lynn Scarlett
      Assistant Secretary for Policy, Management, and Budget

Subject: Protection of Sensitive Information

The events of September 11 and the weeks that have followed should leave no doubts that we are living and working in a greatly changed world. Prominent American icons have been attacked and it is prudent to anticipate that further attacks are entirely probable. Clearly, we must make every reasonable effort to prevent unauthorized or inadvertent disclosure of particularly sensitive information that could aid in planning or executing future attacks.

The current situation requires that we reassess whether certain kinds of information, which may have been available to the public on websites, reading rooms, libraries and similar venues should continue to be available in these ways. It also requires us to reassess the extent to which other information, on internal computer systems and at other locations such as offices, needs to be made inaccessible to the public and even to employees who do not have a legitimate "need to know." The determination of whether information is particularly sensitive requires thoughtful, rational decisions on the part of all senior managers, program managers, and information owners. To determine if certain information is particularly sensitive, at a minimum the following questions should be asked:

- Would the decision to protect or not protect the information be in compliance with public laws, Executive Branch directions, Federal standards, and DOI's own policies? Further, in the absence of precise guidance, would the decision to protect or not protect be considered reasonable and prudent?

- Could a compromise of the information reasonably be expected to directly or indirectly lead to a loss of life, property, money, or public confidence?

- Could the compromise of the information reasonably be expected to prevent or severely hamper DOI's ability to conduct its primary missions or respond to a future attack?

Once a determination is made that the information is particularly sensitive, physical and electronic access should be restricted to only those employees or other authorized individuals with a legitimate need to know the information. At the same time we must recognize that any actions to restrict public and other access to information must be undertaken in accordance with legal requirements. Also, where an individual requests information in accordance with the requirements of the Freedom of Information Act (FOIA) or the Privacy Act, release of the information is required unless protected from release by the applicable statute.

Questions concerning the protection of information should be directed to the Deputy Chief Information Officer on (202) 208-6194. Questions concerning FOIA or Privacy Act should be directed to appropriate bureau FOIA or Privacy Act officers.

CC: Bureau CIO's, Deputy CIO's,
Bureau IT Security Managers